The Geographer at Work

The Geographer at Work

Peter Gould

London and New York

First published 1985 by Routledge & Kegan Paul plc
Reprinted 1989, 1990 by Routledge
11 New Fetter Lane, London EC4P 4EE
Simultaneously published in the USA and Canada
by Routledge
a division of Routledge, Chapman and Hall, Inc.
29 West 35th Street, New York, NY 10001

Set in Linotron Times
by Input Typesetting Ltd, London
and printed in Great Britain
by T. J. Press (Padstow) Ltd.,
Padstow, Cornwall

British Library Cataloguing in Publication Data

Gould, Peter
The geographer at work.
1. Geography
I. Title
910

ISBN 0–415–03674–7

Library of Congress Cataloging in Publication Data

Gould, Peter, 1932–
The geographer at work.
Includes index.
1. Geography – Philosophy.   1. Title
G70.G66   1985   910'.01   84–24961

ISBN 0–415–03674–7 (pbk)

Contents

Contents

Figures

Figures

Figures

Two and two is four
Four and four is eight
Eight plus eight makes sixteen . . .
Once again! says the schoolmaster
Two and two is four . . .
But there is the lyrebird
Flying in the sky . . .
            and the little child calls to it:
Save me, play with me
Little bird!
And the lyrebird lands
And plays with the child
Two and two is four . . .
Once again! says the schoolmaster . . .
            four and four makes eight
Eight plus eight makes sixteen
And sixteen and sixteen, what does *that* make?
Sixteen and sixteen makes *nothing*,
And certainly not thirty-two , . .
And the lyrebird plays
And the little child sings
And the teacher yells:
Stop fooling around!

From Jacques Prévert, 'Page d'écriture', in *Paroles*
(Paris: Editions Gallimard, 1972), pp. 145–6.

Preface
Breaking out of the schoolroom

Once upon a time, the French poet Jacques Prévert wrote a charming poem about a little child sitting in a stuffy schoolroom, listening to a droning schoolmaster, and wishing he could play in the fresh air and sunshine just outside the window. We have all experienced this sort of feeling, for even if we enjoy what we do, there are still times when we feel a bit fidgety and cooped up. Learning is all very well, but occasionally it is fun to break out and climb a tree, or just lie in a meadow watching the clouds go by.

Every once in a while, people in the academic world also like to break out, although not necessarily in quite the same way, or because they are bored. In fact, people in universities are about the least bored you will find anywhere, and most of them love what they are doing – even if many of them are terribly overworked and underpaid. All of them have chosen to work at things they find intellectually worthwhile and exciting, and, more than most people, they still retain some control over their daily and professional lives. Even so, there are times when some of them feel a bit like the small child cooped up in the stuffy classroom, not because they want to do something radically different, but because they would like to talk to different sorts of people, reach out to other types of 'students', and share what they feel are some very exciting stories with a larger and more diverse audience.

In a university, such 'popularizing' activities are usually frowned upon. After all, a scholar is meant to be serious, lead a cloistered life, teach his students, do research, and write up the results for publication in professional journals. Everyone knows that is what *real* scholarship is about; at which point those who have had the true scholastic life revealed to them stomp off like the bassoon grandfather in 'Peter and the Wolf'. But Peter, you will remember, took no notice of his grandfather, and despite the wolves in the forest he opened the garden gate and went out to play in the meadow. Which is precisely what this Peter feels like doing at the moment.

I have had the good fortune to take part in a field that has developed in quite extraordinary ways during the past thirty years, but the exciting story is known only by those inside the garden, the professional geographers who read and write for the professional journals. Full of specialized, and often rather esoteric and mathematical articles, these professional communications seldom provide interesting bedtime reading for a wider audience. But when the articles, research monographs and consulting reports are carefully selected and pieced together, they actually tell a story of explosive development and exciting intellectual renewal. It is difficult to think of any field that is so alive, so full of potential, so capable of engaging many of the important problems that face the human family in its planetary home. Geography is very different today compared to what it was thirty years ago. Few outside the field know what has happened, although many have an instinctive sense that we ought to know, deeply and thoroughly, much more about the relationships between the given and shaped environments and those who do the shaping. No matter how irreverent and tongue-in-cheek the cover of this book seems to be, in the quiet of the night most of us are still awed by the thought of that satellite photograph showing us our gentle blue globe hanging there in the darkness. I have the feeling that these sorts of thoughts, compounded from a strange mixture of curiosity and reverence, have characterized the geographic way of looking from time immemorial.

Every field emerges out of its past, whether in slow, methodical fashion, or in great spurts of explosive change and renewal. And to say this is more than just a tautology: we either build by slow and patient accretion upon what we have, or we renew by reacting against a particular tradition of inquiry. Yet by confirmation or reaction, we still acknowledge the tradition that is always there. In fact, we often do a mixture of both, and the story of geography over the past thirty years tells of great surges of reaction and renewal, as well as continuations of past traditions. Yet in recent years the weight has been on the former, and it is these exciting new developments that shape the major part of the story I want to tell. In selecting these, I have been forced to exclude a number of aspects of contemporary research, tending to include those that seem fresh and new, the sorts of things that make people say, 'Oh, I never knew geographers did that!' Most people have little idea what modern geography is all about, and this book is for them, not the professionals. By definition, the professionals do not need it. So this book is designed to pique your curiosity, not to assuage it. By telling a part of the story, I hope I can give you a glimpse of that workaday world of the modern geographer in all its variety and excitement.

Jacquetta Hawkes, a fine poet and really a geographer at heart, once wrote:

> Snuff out your tapers, ancient Pleiades
> You seven are seven no more but empty light
> Fade fast Andromeda and Cassiopeia;
> From too close looking follows loss of sight.
>
> Jacquetta Hawkes, 'From Too Close Looking',
> in *Symbols and Speculations* (London: The Cresset Press, 1949)

She was writing about someone she loved, and the writing of this book will also disclose a close looking, perhaps one containing more than just a hint of amorous, even prurient, interest. For as we shall see, Geographia is a delicious and seductive wench, and I confess to a passionate love affair with her since the age of 15. A full and ripe woman of many moods, she had made me happy and sad, exasperated and elated, joyful and angry. There have been times when I swear I will have nothing more to do with her, that this time it is all over. But she only has to let her tunic slip to disclose another of her charms and I am smitten all over again.

My wife knows all about this affair, and indulges the liaison with quiet amusement and gentle understanding. To dedicate a book about one's mistress to one's wife is admittedly unusual. But Geographia is quite a lady.

And so is Jo.

Autrans, Le Vercors
May 1984

Acknowledgments: A Posie of Other Men's Flowers

I have gathered a posie of other men's flowers, and only the thread that binds them is my own.

Montaigne

Many years ago, I was introduced to these lines of Montaigne through the title page of a book that is still my favourite anthology of poetry.* But now I know what they mean. Formal acknowledgments can never really thank all the people who helped, with a word here and an idea there, with a map discovered or a diagram sketched, with a phrase coined or a book recommended. The intellectual grist for this particular mill has been very varied indeed, but I hope the loaf is well made. So many colleagues and friends have enriched my professional life over the years that I can never do them all justice.

But for specific items and permissions I do acknowledge with deepest thanks the following:

For photographs: Jorge Gaspar (Henry the Navigator); Horacio Capel (Alexander von Humboldt); the Regional Archives (Herman Krick); Eric Rawstron (Arthur Smailes); Paul Fogelberg (Reino Ajo); Culver Pictures (Ada Lovelace); The Royal Society (Charles Babbage); Torsten Hägerstrand (Walter Christaller); Tilmann Habermas (Jürgen Habermas); André Kilchenmann and Marga Künkele-Lösch (August Lösch); Reszö Laszlo (Edgar Kant); Princeton University Archives (John Stewart); Marjorie Sweeting (Halford Mackinder); Michael Wise (Dudley Stamp); Universiteitsmuseum Utrecht (Christian van Paassen); Noriyuki Sugiura (Joji Ezawa); Nancy Burley (Richard Symanski) and all who contributed personal photographs.

For maps and illustrations: James McClure (page 4); British Museum (2.1); Forrest Pitts (2.2); *Scientific American* (2.3); Armin Wolf (2.4); National Palace Museum, Taipei (2.6);

Acknowledgments

Anonymous Artist I (3.4); Anonymous Artist II (3.5, 28.1); Isobel Robertson (8.4–8.5); Charles Monroe (8.6); Elizabeth Marsh (8.7); Harvey Hutter Inc. (9.1); Prentice-Hall (9.2, 9.8); Sven Godlund (9.4, 13.9); Gerard Rushton (9.5, 23.3–23.4); Brian Berry (9.8); Marie Ciceri and Ronald Eyton (9.9); Howard Gauthier (9.10–9.11); Bernard Marchand (9.12); Thomas Leinbach (10.1); Debra Straussfogel (10.5); Peter Allen (10.6–10.7); Luigi Cavelli-Sforza (11.4); Stuart Fotheringham and the Association of American Geographers (11.6); Andrew Charlesworth (12.1); Barry Riddell (12.2); University of Minnesota Press (12.3); Gunnar Törnqvist (12.9, 13.10, 20.3–20.5); William Warntz and the American Geographical Society (13.2–13.4); Waldo Tobler (2.5, 13.5, 17.10, 17.17, 18.4); Guido Dorigo (13.6); D. Keeble, P. L. Owens, C. Thompson and HMSO (13.8); Richard Eaton (14.1, 20.8); David Ley (17.1); Ronald Eyton and Daniel Grogan (17.3, 18.2–18.3); Susan Evans (17.7–17.8); Cambridge University Press (17.9); Pangiras Michael (17.11–17.12); Michael Goodchild and Mi Yee Kwan (17.15); the US Defense Meteorological Satellite Program (18.1); Roger Agache (18.5); Goodyear Corporation (18.6); William Haxby (18.7); Peter Haggett and Andrew Cliff (19.1, 19.6–19.9); John Wiley (19.2); New Scientist (19.3); Gerald Pyle (19.4–19.5); Reginald Golledge and Nathan Gale (20.1); Cees Eysberg (20.2); David McCann (20.13); William Bunge (21.1–21.4); John Cole, John Beynon and Basil Blackwell (21.6–21.7, 21.9–21.12); Guy Thouvenot (21.8); Graham Chapman (22.1); Annik Rogier (23.5); Lakshman Yapa (23.6, 23.8); Pion Ltd (10.6–10.7, 24.1); Horacio Capel (24.2); Helen Couclelis (25.3–25.4).

For quotations: Editions Gallimard for 'Page d'écriture', and Jacquetta Hawkes for 'From Too Close Looking'.

Finally, I would like to acknowledge with deep appreciation the photographic efforts of Abby Alexander Curtis, far beyond the normal graphic and cartographic duties, the support of Greg Knight who eased the burdens during a difficult period of last minute preparations, and the magnificent typing efforts of Joan Summers back and forth across the Atlantic River.

---

* A. P. Wavell, *Other Men's Flowers* (London: Jonathan Cape, 1944), and many subsequent editions and reprintings. It contains only those poems that the late Field Marshal knew by heart, poems that lifted his spirit and raised his courage during some of the most disheartening and desperate days of the Second World War.

Part I
The geographic explosion

What on earth is geography?

The scene was typical of that extraordinary ritual known as the
Cocktail Party. Across the room, the host was welcoming his
guests with the usual 'So glad you could make it!', and 'Drinks
right over there – do help yourself', while the players in the
game put on their well-practised look of self-assurance, hoping
rather desperately they would recognize someone they knew.
Small knots of people, chatting away with forced vivaciousness,
had already formed into impregnable fortresses, and the channels between them gaped like lonely canyons to shape the drift
of those outside the magic circles of security. Everyone engaged
in earnest conversation or purposeful movement to avoid the
appearance of being disconnected from the rest. A rather
desperate anthropologist, normally quite at home with his
people in the interior highlands of New Guinea, slipped into
his professional participant-observer mode as easily as a soldier
putting on his bullet-proof vest. The slightly anxious, but always
analytical, philosopher realized they were playing Wittgen-
stein's language games, although the fact that his inner security
depended on thinking in a language game himself did not occur
to him at that moment.

Forced from the bar table by the pressure of new arrivals at
the watering hole, and guided to the window by the only available gap in the human maze, I saw her examining the flowers
with what seemed to be professional diligence. 'Roses?' I said
brightly, knowing that I was never terribly good at flowers.
'Peonies,' she replied. 'At least . . . I *think* so.'

'Oh, yes, silly of me . . . it's been a marvellous spring for
flowers.'

'Yes, wonderful. They always seem to do rather well in the
spring.'

'Yes, very good for them, isn't it,' at which point further
conversation about the botanical world seemed unlikely to
enrich our lives.

Groping for something else to fill the silence, she got in her

3

So we ended up parameterizing the nonlinear constraints of the entropy function to ensure that the system trajectory reached a bifurcation point during the numerical evaluation of the stochastically determined boundary conditions . . .

word first. 'And what do *you* do?' she said.

'Oh,' I said, grateful for the usual filler, 'I'm a geographer.' And even as I said it, I felt the safe ground turning into the familiar quagmire. She did not have to ask the next question, but she did anyway.

'A geographer?'

'Er . . . yes, a geographer,' said with that quietly enthusiastic confidence that trips so easily from the tongues of doctors, engineers, airline pilots, truckers, sailors and tramps. After all, everyone knows what they do, and off the conversation goes on the awful 'flu epidemic, the new bridge, the latest jet, the long haul out of Kansas City, the storm in the Bay of Biscay or the doss houses of Saskatoon.But a *geographer*?

'Oh really, a geographer . . . and what *do* geographers do?'

It has happened many times, and it seldom gets better. That awful feeling of desperate foolishness when you, a professional geographer, find yourself incapable of explaining simply and shortly to others what you really do. One could say, 'I look at the world from a spatial perspective, in a sense through spatial spectacles,' or 'Well, actually, I'm a spatial analyst,' both of which would be true up to a point. But such phrases convey no meaning to most people, and leave them suspecting that you need a new oculist, or perhaps an analyst of a different sort.

Or there is the concrete example approach. 'Well, at the moment we're calibrating an entropy-maximizing model for a

4

journey-to-work study in Bogota,' or perhaps 'We're seeing if we can improve on the spatial auto-correlation function as a filter to clean up the pixels', or possibly 'We're using a part stochastic, part deterministic, computer simulation model to examine the threshold values in a regional development programme,' all of which would be equally true up to a point. But the words, with their precise meaning for geographers, convey nothing to others, and end up sounding like some private and deliberately obfuscating jargon. Which would also be true. Up to a point. Often, in a desperate attempt to build a bridge with more familiar words, one ends up by saying, 'Well, actually, I teach geography.'

'Oh really?', and laughing. 'What's the capital of North Dakota?'

It does not have to be North Dakota: it could be the coalfields of Yorkshire, the longest river in the world, the climate of Perth, Australia, the population of India, or the major exports of Zaire – although no one knows enough these days even to think about Zaire. Except geographers, of course. Geographers know all about where things are, and why they are there, and that is what geography is all about, right?

'Right. I remember we did South America in the sixth grade.'

There is no parody here. This sort of conversation reflects most people's vision of what geography is and what geographers do. And in a sense this vision is a quarter right. We do have the responsibility of teaching children about the world they are growing up in, in just the same way that we have the responsibility to teach them their language, mathematics and a sense of their historical and artistic heritage. Thrown into a world not of their own making, how do we help children to make sense out of it except by conveying to them our own sense of word, number, beauty, time and space?

And *space*! Not that outer space of science fiction that is slowly opening up to our probing instruments, but the right-here-at-home space, the geographic space of our small planetary home. We speak so glibly of our shrinking globe, of the growing impact of one culture upon another, of the increasing interdependence of the human family. Events in one part of the world have a direct and immediate impact on another, but too few children – and ultimately that means too few adults – even possess the most elementary information about the world stage, let alone about the human players who produce a constantly changing kaleidoscope of cities and settlements, roads and regions, conflict and cooperation. Let El Niño, the ocean current off Chile and Peru, change its course, and it is catastrophe for the commercial fisheries, a catastrophe that seems to be linked to droughts in Australia 10,000 miles away across the Pacific Ocean. And when three major volcanoes all go off within a year of each other, we all wonder what the effect will be upon our climate and the crop yields in the years to come. Countries like Kuwait and Bahrain have ever-swelling coffers from their OPEC oil, but the countries who feel the pinch the hardest are the poor Third World countries like Tanzania, Guyana and Bangladesh. To keep their trucks and produce moving, they beggar themselves to pay for oil in hard currencies, and that always means less fertilizer, fewer schools, fewer hospitals, fewer roads – fewer everything that might make up a national development plan.

But what do such statements about our physical and human world mean – what 'comes to mind' – if you do not know, if you cannot 'picture', Chile, Peru, Australia, Kuwait, Saudi Arabia, Tanzania, Guyana and Bangladesh? Yes, geographers do have the responsibility for teaching about the places and the spaces, the rivers and the mountains, the structures and connections, and all the interrelationships of the world we live in. Otherwise, how can our world make any sense and have any meaning? But I said that this traditional, but still vitally important, task of elementary teaching was only a quarter of the story. The other three-quarters is going on today in the universities and research establishments of many countries, in consulting firms and businesses, in urban, regional and national planning offices, in government agencies and in supra-governmental institutions like the World Health Organization and the World Bank. One of the outstanding characteristics of the explosive and exciting developments of the past thirty years is the way geographic expertise – that spatial perspective – is informing and illuminating problem after problem over a wide spectrum of concern.

The range is staggeringly wide. Geographers have done fundamental research on helping mentally handicapped people find their way about complex urban areas, on providing more readily accessible medical care in rural areas, on estimating crop yields from satellite photographs, on planning new agricultural regions, on complex problems of urban transportation, on sensitizing potential administrators in the World Bank to the grass roots difficulties facing poor Asian farmers, on reapportioning electoral districts to insure the democratic slogan of 'one person, one vote', on the creation of tourist images in an industry that is one of the fastest growing in the world, on international flows of television programmes, on measles epidemics . . . and so, almost endlessly, on. Most people's reaction to such topics is, 'But I never realized that geographers did *those* sorts of things.'

But they do – and more. And we shall meet many of them throughout this book. Many of the theoretical developments and practical applications are a result of the dramatic changes in geographic training and research of the past thirty years, helped in large part by the exciting parallel developments and changes in our ability to observe and handle huge quantities of information with the increased prosthetic power of the modern computer. Geography, in some of its most fundamental ways, is a very old discipline, and we shall look more closely at some of these origins later. It is also, quite paradoxically, a very new discipline, only represented in the universities slowly over the past hundred years. Of that century, the past thirty years have been the most exciting, and this is the story I want to tell you about in this book.

So that the next time someone says to me at a cocktail party:

'Oh, you're a geographer are you? Er . . . what exactly do you *do*?', I am going to reply:

'I'm *so* glad you asked me that question. I just happen to have a little book on me . . .', and I will reach into my pocket and hand them *The Geographer at Work*.

And perhaps in this way the glad tidings will spread.

Even to those living in the capital of North Dakota.

NORTH DAKOTA

Bismark

7

2 The old roots of geographic curiosity

Right from the beginning geography presents us with a paradox. At one and the same time, it is among the oldest areas of human inquiry in the world, and yet it is also one of the most recent intellectual disciplines if we judge it in modern academic terms. What on earth can we make of this oldest-newest business? We have two clues to help us understand its ancient origins. The first comes from a very simple and direct appeal to our own experience as human beings. Whatever country or world of culture we happen to be born into – Swaziland or Syria, Basque or Welsh – we seem to acquire, in an almost definitive sense of being human, what we might call a primordial sense of time and space. Not the precise modern clocktime or measured space of the scientist or surveyor, but the very simple and intuitive sense that one event comes before another, and that places and things are close or far away. We develop a sense of place from our familiar and comfortable surroundings, yet we also acquire a sense of a larger space, for we can see a horizon and wonder what lies beyond it. This sense of space, with its embedded sense of home and place, seems to be acquired at a remarkably early age, and our experience of it is shaped by language – as is all our thinking. In our everyday speech we use many geographic and spatial expressions as a matter of course, talking about 'regions' of inquiry, 'mapping' out our plans, or saying someone we know appeared very 'distant' this morning. Even the slang expression from the 1960s, 'Hey man, that's far out!', gives us a vivid image of something exciting or different precisely because it is not a part of our everyday world of things close at hand. Our intuitive awareness of the geographic space 'out there' seems to be part of what it is to be human.

But the second clue about this ancient-modern paradox lies in the historical origins of our modern field of inquiry, and the way these origins seem to spring from the reflections of a particular group of people – the Greeks. This is not just a token

8

obeisance, a sort of traditional tipping of the academic hat
before we get on to more important things. The Greek experience was absolutely fundamental, and it is in these old Greek
roots that the modern scientific origins of modern geography
lie. Not that Geographia sprang suddenly and fully computer-
ized from the head of Zeus, because with little scraps of
evidence here and there we can discern even earlier origins.
The evidence, rather importantly, lies in maps: a map of the
world impressed by a scribe's stylus on a Babylonic clay tablet,
an intriguing local map incised upon a flat rock in Bresicia,
northern Italy, and a map engraved upon a piece of mammoth
bone. The Babylonic map of the world, found in excavations
at Sippar, is from the seventh or sixth century BC (Figure 2.1),

Figure 2.1: The
Babylonic 'map of the
world'.

but the cuneiform writing records that it was copied from an
even older tablet. The world is a disc, with mountains, marshes
and rivers, all surrounded by the bitter river – presumably the
Salt Sea. The Bresicia map (Figure 2.2) is less ambitious, but
no less illuminating of the human propensity to record the
world in maps. Scratched on a flat, moss-covered rock overlooking the Carmonica Valley, we have this marvellous *graphic*
expression of a tiny part of our *geo*. Someone more than 4,000
years ago sat there above the valley and incised what looks to
us like a permanent village, laying out the houses and
connecting paths, as well as some of the fields and animals.
There is even evidence from the geometric course of the
stream, and the round wells, that irrigation was used to grow

Figure 2.2: The *Camuni* map from Brescia, northern Italy, incised on a flat rock overlooking the Camonica Valley. Dated at 2,000 B.C., it actually overlays still earlier maps that are too difficult to discern clearly today.

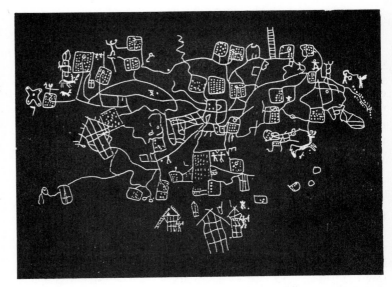

crops. As for the map scratched on a piece of mammoth bone (Figure 2.3), it was found at an ice age camp site on the river Dneper that was a living human community 15,000 years ago. It is the oldest map in the world, and the very first piece of evidence we have for geographic thought and imagination. Who these people were, what they did and thought, we shall never know. But they left behind the evidence that they too shared our modern sense of map-making, with all its primordial concern and wonder for spatial relations.

From other maps and records, we feel sure that many field

Figure 2.3: The oldest map in the world, carved on a piece of mammoth bone.

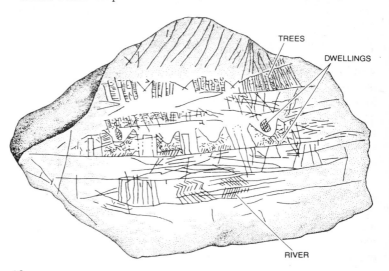

TREES

DWELLINGS

RIVER

and property boundaries in the ancient world were measured according to carefully recorded *geometrical* principles. Geometry and geo-graphy, measuring the earth and writing about the earth – these two modern fields were hardly separated in those days. This sense of wholeness, of knowledge all connected up and part of a unity, is precisely what comes to us when we reflect upon our Greek heritage that itself drew upon the worlds of Crete and Babylon and Egypt. When we reach back to those times, and try to retrieve a sense of the . . . well, what shall we call it? The intellectual *milieu*? No, too modern, French, and redolent of Left Bank cafés. Perhaps the *Weltanschauung*? Oh no, much too heavy and Germanic for the Greeks! How about the *immediate thinking life*? Yes, that's better. When we reach back and try to recapture the immediate thinking life of the Greeks, we must never impose our own chopped up, partitioned and compartmentalized world of the late twentieth century. Our own world is splintered and fragmented into those little nineteenth-century boxes labelled biology, physics, sociology, economics . . . and, if the gods do not strike us dead, *education*. But these modern labels and Johnny-come-latelies do not help us make sense out of the Greek world where the origins of geographic science lie. Indeed, they only force an alien and distorting framework upon it.

We have to place ourselves back in a world in which even the field of *logic* – that most distinctively Greek-rooted word – was only just emerging with the meaning we give it today, and we have to listen to Greek words very carefully. When Herodotus, perhaps the father of geography, wrote his *Historia* around 480 BC, he was not writing *history* as we know it today. History and geography had yet to emerge as separate fields, for *historia* means to explore (very geographic!), or to make visible, and perhaps the best meaning we can give it today is the word *inquiry* itself. And inquire Herodotus did, often from first-hand experience, about earthquakes and cities, the depths of the sea and delta-forming rivers, ships and how to build them, climates and mountains, all mixed up with the colour of human custom and local culture. A native of Caria, at the centre of that hotbed of explosive intellectual life in the eastern Aegean, he travelled widely over the known world in the course of gathering the materials for his book. As a highly educated man of his times he had a fine model, one whose own works were so intimately known to him that they must have formed part of his very being. For the *Iliad* and *Odyssey* of Homer are much more than great epic poems about a mythological past.

Armin Wolf
Max Planck Institute for European Legal History
1935–

11

When Schliemann uncovered the city of Troy in 1871 – layer upon layer of Troys – the modern world realized that more than just stories were there. Today, after seventy previous attempts by others, the German classical scholars Hans-Helmut and Armin Wolf (who can only be geographers at heart) have done the same sort of thing for the *Odyssey*, reconstructing with a truly marvellous display of patiently gathered evidence the actual journey of Odysseus on his long roundabout way home to Ithaca (Figure 2.4). Piece by piece the text is matched with the geography, so that even Homer's references to Dipylon shields allow us to retrieve the crossing point of Odysseus across the toe of Italy. The *Odyssey* is a geographical *historia* too. So do geographers claim Homer as one of their own? Well, we claim the best of everything else, and since Homer belongs to all of us, why not?

The point, of course, is that knowledge and inquiry seemed to have a coherence in those days that has never been seen since, and the names that come down to us represent a great flowering of the human spirit, one that would not really occur again until the Renaissance. Two hundred years after Herodotus, Aristarchus of Samos would hypothesize that the 'fixed stars and the sun remain unmoved, that the earth revolves around the sun in the circumference of a circle'. The words are Archimedes', and nearly 2,000 years later, Copernicus would gratefully acknowledge his debt. Building upon this startling idea Eratosthenes, the librarian of the great Library of Alexandria, would measure the circumference of the earth, and his *Geography* not only explained his mathematical methods, but provided the foundations for the great *Geography* of Ptolemy, the father of cartography who located 8,000 places on the

Figure 2.4: The journey of Odysseus, reconstructed by Hans-Helmut and Armin Wolf in *Die wirkliche Reise des Odysseus*, and somewhat simplified in this redrawn English version.

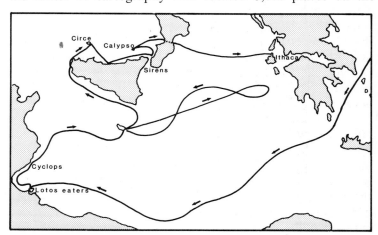

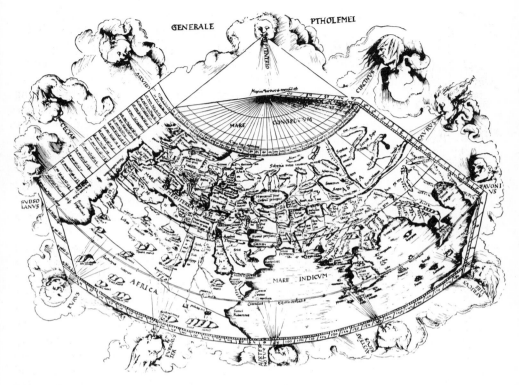

Figure 2.5: The Ptolemy map of the world, the first based on a coordinate system to record and locate the 8,000 places known in his day. This version was printed from a woodcut by Lienhart Holle at Ulm in 1482.

surface of the earth using the same coordinate system of latitude and longitude we use today (Figure 2.5). For this he had to develop what we know now as trigonometry, as well as extensions to the algebra of his day, so we see here the close liaison, the almost total intertwining, of geography, mathematics and astronomy. It is not just a question of the similarity between the two words, geometry-geography, but an ancient and mutually supporting partnership between fields that at one time were never really separated.

Unfortunately, the marvellous intellectual and artistic efflorescence of the Classical and Alexandrian Greek worlds could not last. The Roman spirit was of a different order, and fundamentally inimical to the Greek tradition of probing inquiry, the sort of inquiry that was prepared to go wherever the questions might lead. This same spirit, the same pushing of the questions to the limits no matter what the consequences, is precisely the driving force of our own times – for good and

13

for ill. But it would take the two millennia between us and that Greek world to bring it back. When the Romans set fire to the ships in Alexandria harbour in 47 BC, the flames spread to the great Library, consuming almost all of the 400,000 manuscripts. Later on, the temple of Serapis, used as a vast annex to hold 300,000 more, was destroyed by the Christians when the Emperor Theodosius banned all 'pagan' religions. What few texts remained were nearly all destroyed by the new religion of Islam as it swept the Arab world from India to West Africa in the seventh century AD. For six months the baths of Alexandria were heated by crackling parchments covered with the scribbles of the unchosen. The absolute truth of the Bible and Koran were all that were needed by the two worlds that called each other heathen and infidel, but today the soul of the unbeliever numbs at the thought of the precious human heritage destroyed in the name of religious truth. For a thousand years, Europe slept content with its meagre prescribed texts that few could read, and contemplated a flat earth placed by God firmly at the centre of His universe. It was the geocentric universe of Ptolemy, a way of looking that overrode the heliocentric view of Aristarchus of Samos. If geographers claim the best of ancient inquiry, we also have to have the courage to acknowledge responsibility for some of the greatest mistakes.

In the Arab world, after the initial excesses so typical of men flushed by victory, the scientific knowledge of the Alexandrian Greeks was absorbed and in certain areas extended. For reasons that are not entirely clear, the world of Islam did not restrict scientific and mathematical inquiry, and at centres of learning like Cairo, Algiers and Timbuktoo various combinations of geography, astronomy and mathematics flourished. Al Khorizmi, one of many distinguished Arab scholars, made considerable contributions to both the algebra and geography of his day, once again combining and keeping intact the old liaison between mathematics and geography. In our computer age, we acknowledge him daily, for he gave his name to our modern word *algorithm*, meaning a step-by-step procedure that leads to a desired answer. He also reflected deeply on some of the relationships between the size and spacing of towns, working from simplified geometrical principles in an attempt to discover some degree of rationality in the pattern and structure of human settlements. We shall see later that this anticipates a fundamental theoretical theme in contemporary geography. The geographic knowledge of the Arab world, set down in a formal and systematic fashion, was considerable, if for no other reason than that the empire itself eventually stretched in a great

14

15,000-kilometre arc from Mauritania in the west to Indonesia in the east, all united under a common religion, language and script. Long pilgrimages were made to Mecca from the outermost boundaries, and when the Emperor of Mali made his pilgrimage with an enormous caravan of camels across the Sahara, he returned via Cairo and the important towns and universities of North Africa, depressing the price of gold for a decade with his lavish purchases.

The extension of geographic knowledge by the Arab world, and the preservation of the few remaining Greek texts by the Byzantine Empire at the eastern end of the Mediterranean, finally provided the two sources for the re-entry of such knowledge into the virtually closed and intellectually stagnant world of medieval Europe. This is not the place to attempt an outline of the causes and preconditions of the Renaissance, but two events are worth pointing to: the flood of Greek scholars to western Europe fleeing ahead of the conquering Turks in 1453, and the beginnings of what we now call the Age of Exploration. In a sense, the scholars brought to western Europe the ideas of a past time, while the sailors carried news of a new and expanding space. There was an explosion of geographic knowledge. In the fifty years 1472 to 1522, the geographic 'state of the art' went from a belief that any fool sailing too far would fall off the edge of the world disc, to the complete circumnavigation of a round earth by the sailors of Magellan. In the meantime, the tiny caravels of Henry the Navigator had rounded the western bulge of Africa to found El Mina (and still today Elmina) in Ghana (1482); Columbus did not fall off the edge, but bumped into the Americas by mistake (1492); and Vasco da Gama sailed around the Cape of Good Hope and into the Indian Ocean (1497). Europe began to be flooded with new accounts of strange lands, strange animals, strange plants and strange and different people. With the old Greek spirit of open inquiry permeating intellectual life from the translations of Greek and Arab texts, with the hammer blows of new information about a larger world disclosed by the new technology of sailing, clocks and other navigational instruments, and with accounts of all the new discoveries made widely available by the printing press, the old closed world of medieval Europe broke open to disclose the Renaissance.

They were brave men who opened up that European world to the outside, braver perhaps than the first astronauts who sailed 500 years later in a different sort of ship to the moon. When you drive today to the high cliffs of Sagres, the extreme southwestern point of Portugal where Henry built his great

Henry the Navigator (from an azulejo tile portrait) 1394–1460

school of cartography and navigation, you leave behind the warm sunny world of the Algarve, with its grapes and olives and almond trees, and enter a harsh tundra-like landscape shaped by a howling wind from the cold Canary current, with a clear reach of 8,000 miles to the pack ice of Antarctica. It took men of great courage, with high orders of seamanship and navigational skill, to brave those grey roiling waters. The geographical knowledge we have today was pieced together at great human cost, and with our modern technological arrogance we forget too quickly.

And speaking of arrogance, we should never forget that we are looking at these events purely from the European point of view. The Arabs in their dhows already knew the coastal waters of the Indian Ocean like the back of their hands, and the Chinese were masterful sailors. Chinese habits of courtesy required them to provide an escort home to those who had presented gifts to the court, and it was this obligation that eventually led them to sail to the coast of East Africa in their great seagoing junks. One silk painting (Figure 2.6) shows a

Figure 2.6: A giraffe from East Africa presented to the Emperor of China.

giraffe presented to the Emperor, so we know of direct ties between these parts of the non-European world. In the Pacific, the great Polynesian migrations had already taken place, and we know these people were superb navigators. Delicate 'maps', made from the spines of coconut palms and shells, provided a geometric lattice that guided canoe voyages over vast distances.

16

The awareness of the horizon, and the desire to see what is beyond, seems to be shared by us all.

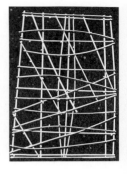

Nevertheless, geographical knowledge consists of much more than a sailor's logbook after a voyage. The explorers by sea certainly limned the coastlines, but the maps being pieced together often looked like thin-skinned balloons – a bit soft and distorted on the outside, and really hot air in the middle. The vast interiors of the blank continents were mostly spaces in which the early cartographers could exercise their artistic licence, spiced by their often vivid imaginations. Exploration over the land was more difficult than over the sea, hard facts about the interiors were difficult to come by, and we tend to forget that it is only just over a hundred years ago that we finally discovered where the source(s) of the Nile lie. It is only now that satellite photographs are allowing us to map the deep interior of Brazil, and they disclose huge tributaries of the Amazon river we never even suspected before.

At the same time, those blank and rather embarrassing spaces were also seen as invitations to fill them. A second phase of systematic exploration of the interiors began, and the reports of these new lands were often accompanied by beautiful illustrations of the plants, animals and cultural artifacts of the local people. At the end of the eighteenth century, Alexander von Humboldt's descriptions of physical landscape and human society were typical of many, but he also began to reflect on those two-way relationships between human beings and their environments – not just the way people modify their natural landscapes, something easy enough to see in Europe, but how the physical setting of climate, soil and terrain alters them in turn. Pushed to extremes, these ideas were to become the doctrine of Environmental Determinism, the notion that we are almost wholly shaped by the physical environment. This was an idea to be pushed hard even into the beginning of the twentieth century, when the pendulum swung dramatically the opposite way, and Determinism was softened first to Probabilism and then to Possibilism, until eventually environmental influences were almost denied completely. This position, of course, is as extreme as the most rabid Determinism, because it is simply unreasonable to deny *any* influence of the physical world upon us. We have all experienced hot muggy days when we feel utterly lethargic, and wonderful clear and crisp mornings when we feel we could do anything. There is a strong correlation between skin colour and latitude, and the rapidly growing incidence of skin cancers among people of northern European descent points to the selective mechanism of protec-

Alexander von Humboldt
Independent Scholar
1769–1859

17

tion. Natural trace elements in soil and water directly affect our health, and we have maps of goitres showing that both the alpine recruits for Napoleon's army, and modern-day settlers in Texas, need small but vital quantities of iodine. Fluoride treatment for teeth was discovered by maps showing areas of low dental decay where there was a high natural content of fluorine in the water . . . and so on. To say we are unaffected by the physical world is nonsense, and denied every time we foul our own nest with cancer-producing chemicals and the debris from atomic plants and explosions.

Part of the desire to fill those blanks on the maps came from that old Greek curiosity to *know*, but it was often strongly aided and abetted by the overwhelming political fact of European colonial expansion into other parts of the world. South America was divided early on by papal decree into two parts to be administered by Spain and Portugal, and later Africa was chopped to pieces by drawing blue lines across the blank maps in the chancellories of Europe. Britain, France, Germany, Italy, Spain and Portugal (and others like Denmark and Brandenburg at various times), all engaged in the race, and sometimes when their surveying teams saw each other on the horizon, they literally raced each other to the nearest unsuspecting village to be the first to plant their nation's flags. Naturally, when you have established vast claims to control and administer a huge chunk of blank map, you have to find out what is there. And who better to find out than geographers and map-makers? Thus was the old root of Greek curiosity entwined with the political power that comes from modern, systematic knowing, and it is no accident that geography really first appears as a separate discipline in European schools and universities at about the same time as imperial ambitions were becoming more and more rapacious. Knowledge was power in a very direct and immediate sense. It was not a very honourable beginning to modern geography, and it was certainly not the only cause, but it is there nevertheless.

The geography of colonial expansion tended, quite naturally, towards a systematic recording of what was where. It was concerned initially with making an inventory of things at particular places, a perfectly good task if you really do not know and have some purpose. It is a tradition that continues today in the Soviet Union, where many young geographers just out of university serve their country and 'pay off' their university education by compiling careful accounts of natural resources in vast areas of that country that have not been systematically explored on the ground. The problem was that the 'inventory' tradition soon became the 'research' tradition in many of the newly emerging departments of geography in the universities, and respectable and challenging intellectual reasons had to be cobbled up to justify the expenditures of all the time, money and energy. These were found in the idea of the region, and what was termed 'geographic synthesis', the job of putting it all together, and showing how everything on the inventory list – the rocks, soil, vegetation, climate, agriculture, mining, manufacturing – was interrelated. Out of these relationships emerged a region's own character, even its personality, and it was the geographer's task to unravel and explain such complexity.

Unravelling and coming to a deeper understanding of something as complicated as human-environmental relationships in a region is a daunting intellectual task, and usually it did not work. Too often doing geography became doing a checklist, with geographers arguing in their journals about the items that should be included and which were the most important. You started at the top of the list, with the rocks, soils and so on, and when you got to the bottom you were finished. So, nearly, was geography. A parody of the early 1960s contained an article on *The Geography of Willow Creek* by Herman Krick:

> The Willow Creek Region is the fifth largest region in Willow Creek County. Though it ranks only as the 94th largest region in Phydo's *Natural Regions I Have Known*, it is the 92nd largest in its state. . . . There are many characteristics which distinguish the Willow Creek region from other regions, but none is so significant as location. No other region is located where Willow Creek is located. . . . Not very much of the region is located at the lower end of the creek, and it is safe to say that its geography is up the creek.

In the next issue, an *In Memoriam* recorded 'with the greatest sorrow' the death of Herman Krick (1869–1963) 'at the age of 94, only two years before he was due to retire from active research'.

> Although, like all regional geographers, he was an expert in every field of geographical enquiry, he will be remembered and revered by future generations for his pioneering work on the Willow Creek Region. Starting with his doctoral dissertation . . . subsequently translated into German and expanded by footnotes into a four volume work . . . he spent his life meticulously sub-dividing the macro-Willow-Creek-Region and categorizing the sub-divisions. His six books on the Willow Creek Region, and more than three hundred research papers, will stand as a perpetual monument to Regional Geography.

Herman Krick
Flotsam & Jetsam University
1869–1963

Sometimes 'land use' became the rallying theme, and detailed maps were compiled showing how bits of the land were used, often recording the information in esoteric numerical codes

19

that seemed to make the whole thing much more 'scientific'. I can remember a geography field camp in the 1950s when about twenty of us were scattered over the agricultural landscape of southwestern Wisconsin with soil augers, plane tables and air photos clutched in our hot little hands. Periodically, we were meant to plunge our auger into the soil, determine the type, record the slope, the crops, the vegetation and so on, and mark in each field or 'natural area' with a complicated fractional code. Every evening, after a day in the blazing sun and 42 degrees centigrade in the shade, we were collected and taken back to the geographic 'operations room', where we added our day's information to the master map.

Of course, as students we only looked stupid: after the first day of almost total dehydration, we quickly dragged ourselves out of sight over the brow of the nearest hill, found a good vantage point in the shade, and filled in the tracing paper taped over the air photos pinned to our plane tables by making judicious 'estimates'. It was, of course, totally dishonest intellectually, but I confess it here as the only rational response to a blatantly inane piece of busy work. After two weeks, the master map was 'done', and that was that. No use was ever made of it, and apparently none had been foreseen from the beginning. The futility of the whole task was only equalled by the next one: to compile a land use map of an urban area, with its lawyers' offices, gas stations, drug stores . . . in order to find the Central Business District, known to professionals as the CBD. Ordinary people called it *downtown*, and any 5-year-old could have told you where it was without a land use map.

Something had happened to geography. Rightly or wrongly, the regional inventory and land use map became the hallmark of what geographers did, although some might specialize in some items of the list all over the place (systematic geography), while others became experts in all the items in a few places (regional geography). Certainly this image of geography became prevalent outside the profession, both among university colleagues and people in other walks of life. Few people bothered to ask what geographers did, because most people knew: geographers recorded facts about places, and you learnt about these in grade school. In the United States, geography in the schools changed from a fundamental subject in the nineteenth century to a part of the pablum-like mishmash called social studies in the twentieth. In most European countries, geography was still taught as a separate subject in schools, often over eight or nine years, but the main purpose of the university departments was to train the schoolteachers who

20

would teach the children – the nineteenth-century tradition continued. Fair enough, no one would wish to deny that teaching children about their world is a very important task. In many countries of the world, the first geography books – even those at a very simple level for schoolchildren – have only been written quite recently. Before Dudley Stamp started Burma's first geography department at Rangoon in the 1920s, there was no *Geography of Burma*, and Ernest Boateng's *Geography of Ghana* was only published in 1959, and has not been kept up to date. The story is the same in many Third World countries, and many of the 'histories', in the sense of a systematic, published account for children and adults, are equally recent – meaning that they have only appeared during the last few years.

But these elementary, factual accounts were not good enough. Another generation was coming along attracted to geography because it focused upon that primordial spatial dimension in human existence, a dimension that has assumed much greater prominence in the last few decades as we try to plan with greater care the humane and decent use of geographic space, repair some of the foulness and damage we have done, and think through the use of precious renewable and non-renewable resources. Underpinning such modern concern are the old themes of geographic inquiry: a fundamental curiosity for what lies on the other side of the hill, a fascination for the interrelationships between human beings and their physical environments, and a sense of the interconnectedness of things – both human and physical – in a regional setting. But these old themes were to be examined from different and newly emerging perspectives, and in the process geography underwent a jolting (r)evolution. Let us see what happened.

3 The (r)evolution in geography

Unless change is truly catastrophic, the labels *revolution* or *evolution* are employed very much as a matter of taste when it comes to describing developments in a field. The 'Old Guard' (some of whom can be distressingly young), tend to use the term *evolution*, emphasizing that all this new-fangled stuff is hardly new at all, nothing more than a small outgrowth of what went before, and anyway we were doing it all along. And, of course, in some ways they are often right. The 'Young Turks' (some of whom can be delightfully old), tend to use the term *revolution* when talking about the changes they helped to bring about, emphasizing the tremendous differences between the old way of looking at things and the new, modern, contemporary, fresh and, *obviously*, intellectually more profound way we have today. And, of course, in some ways they are right, too. Which is about as close as you can get as an author to sitting astride a fence.

The problem with such fence-sitting is that the tops of fences tend to be a bit sharp and narrow, and eventually you are forced to swing your legs over to one side or the other. So I am going to swing my legs over to the side of *revolution*, but you must realize that the motive is mainly one of comfort, rather than total conviction. Hence the (r)evolution in the title, and I am not going to throw a tantrum if others want to toss the (r) away.

Nevertheless, evolution or revolution, something very dramatic happened to the old field of geography in the late 1950s, and it has been happening ever since. Perhaps the most convincing evidence for this change appears in the way geographers have been using some particular words over the past thirty years. After all, we write and speak in language, and, in a more important and fundamental sense, we *think* in language. If you do not believe me, try thinking not in language, without 'having a conversation with yourself'. Talking to yourself is not a sign of madness, but a sign of normal thinking, a 'Let's see, what am I going to do now . . .', or 'I wonder what would happen if . . .'. Asleep or awake, we talk to ourselves, so our words ought to tell us something about what we are thinking. So it is in a field like geography: the words we use in our professional 'conversations' tend to reveal what we are thinking about at a particular time.

So let us take a look at some words geographers have been using recently, words to be found in the titles of their professional conversations, the journal articles they write to carry their ideas to each other. In particular, we are going to follow the six words (and their common variants) – *spatial, theory, model, regional, structure* and *planning* – over the last thirty years in 21 professional geographic journals, or journals where geographers regularly publish. I do not guarantee *complete* accuracy, and students in a recent seminar of mine will attest that sometimes your eyes start to glaze over when you are scanning 6–8,000 titles, but the frequencies of the words are about right.

Now do you see why I swung my legs over to the *revolution* side? Look at the way some of those words have exploded in the last few decades, especially *spatial, model* and *regional* (Figure 3.1). In the early 1950s, the word *spatial* hardly ever

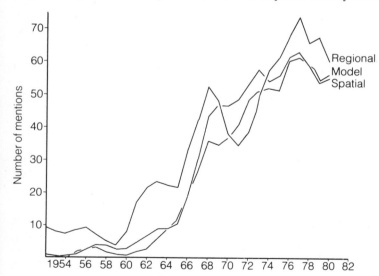

Figure 3.1: The frequency with which the words *spatial, model* and *regional* have been used in the titles of geographic articles over the past 30 years.

appeared in geographic discourse, while today roughly 50 or 60 professional articles employ it in their titles every year. I think we can get some sense of the meaning of this if we trace that 'spatial' line on the graph: it starts from nowhere, gradually picks up speed into the early and middle 1960s, and then explodes in the late 1960s to early 1970s, the same years as other 'revolutionary' activity in the universities. These were the years of the anti-Vietnam War movements in America (Kent State, Berkeley, etc.), and the years of peace marches and other protests against Establishment policies (the Sorbonne, etc.). The curve assumes very roughly an elongated S-shaped form – slow start, rapid rise, slow levelling off – and later on we are going to find that this is typical of a new idea or innovation diffusing through a community over a period of time. The question is why would a word like *spatial* diffuse so explosively through a community of geographers?

I think there are several reasons. First of all, the word itself obviously plucked a deep chord in many of the people forming a new generation of geographers. Speaking for myself, the word resonated deep down inside, and I have the strong suspicion it was a resonance 'in tune' with that old, primordial sense of space we all have at various heights of awareness – the sense of close-to and far-away curiosity about the patterns and structures we see on maps. It is a strange thing, but many professional geographers will tell you that as young children they spent hours poring over maps and atlases, wondering about the shapes and rivers and seas and mountains . . and thinking about what might be 'there', just over the horizon.

But it was also more than that. *Spatial* became an identity word, a word that said, 'I look at things differently from the old geographic way,' a word that gave a generation a sense of breaking away and doing something different. 'Well why not use *geographic*,' said some, 'after all it means essentially the same thing, and it was good enough for your grandfather.' And that was quite true: *spatial*, as it was commonly used, did more or less mean the same thing as *geographic*, except that *geographic* had an old-fashioned, rather prosaic feel to it, and you always had that awful problem of explaining what it meant exactly – the regional checklist and the capital of North Dakota all over again. *Spatial* was new, it sounded much more scientific, and people were so impressed they did not usually ask you what it meant. There was certainly a bit of jargon in it, no question about it, but it also pointed to new problems and very different ways of thinking about them. As an adjective, it was often linked to such nouns as *organization*, *analysis*, *structure* and *relationship*, and all of these expressions had a very different feel to them than the regional inventories and checklists, or the maps of land use which no one knew quite what to do with. Part of the rise of the word *spatial*, of course, is a result of an almost equally dramatic rise in the journals them-selves. Only seven of the 21 we are considering existed in 1953, but this rapid growth is only one more indication of the explosion that occurred in the 1960s (six new journals suddenly appeared in the years 1967–70), and remember we are only looking at journals in English from the United States, Britain and Canada. Elsewhere around the world, and lagging only slightly behind, we see the same growth in geographic journals and those of related areas. We shall look more closely at some of the reasons for this growth in a minute.

It was also in the late 1960s that the use of the word *model* began to grow rapidly,and now we are dealing with a term that really points to some fundamental changes in the way geographic inquiry was carried out (Figure 3.1). Again we have that long S-shaped line from nothing in the 1950s to the high frequencies today, with the steepest rise in the late 1960s. What are the deeper implications of all this? Well, when you begin an investigation of something, say the way patterns of land use form in a city, and you start by constructing a *model*, it means that your thinking is directed away from a particular city at a particular time, and you try to capture in your model the essen-tial things that explain why patterns of land use in many cities all seem to exhibit deeper similarities. If you are clever at this process of abstracting the essence of land use patterns, you can then turn around and use your model as a sort of simplifying lens to look at all sorts of different cities, and so make more sense out of what initially appears to be a chaotic patchwork quilt of land use types – commercial, residential and so on. This idea of a geographic model is so important that I think we ought to look at a specific example, and see what this process of simplifying and abstracting really means.

Instead of making a three-dimensional scale model of a city, in the way an architect or town planner might, we are going to simplify everything down to a two-dimensional graph. You can imagine taking a slice right through the centre of a city, from downtown out to the suburbs, so that the horizontal axis on the graph is simply the distance from the centre. On the vertical axis we are going to record the price of a hectare of land, and this means immediately that we have brought human beings into the picture. The price of a block or hectare of land in various parts of the city will depend on what people are willing to pay for it. Thinking in these purely economic terms for the moment, what they are willing to pay for it will depend on how much rent or return that piece of land will generate for them when they use it for a particular purpose – for shops, hotels,

25

apartments, warehouses, residential bungalows and so on. Of course, land use in a city is shaped by all sorts of other things than just rent – for example, perfectly good political decisions that result in zoning laws – but I am deliberately trying to keep this first model very simple. It is characteristic of constructing geographic models that you make them as simple as possible at first, and then, if they seem to help you to come to some better understanding of what is going on, you can always make them more complicated later on.

Let us see what might happen in our abstract model city, and in particular what sorts of things we might expect to find downtown close to the centre. Because the centre is literally 'in the middle of things', it is the most accessible location and therefore seen by people as highly desirable. At least until traffic congestion and the fumes of automobile exhausts drive people away: I sometimes think the streets of New York, London and Paris are going to grind and screech one day into congealed ribbons of metal, and archaeologists thousands of years from now will be scratching their heads over these strange artifacts. If the centre is seen by people as the most desirable location, they are going to pay lots of money for those relatively few hectares sitting downtown. That means that the things they build there will have to produce a big profit *per hectare* (remember this is a *geographic* model, and space and distance are on both axes of our graph), and these things will be . . . what? Well, the things we find so often downtown in a modern city: office buildings of rich businesses, rather exclusive and snobby shops, and expensive restaurants, often piled on top of each other, sometimes 120 stories high (for example, the World Trade Center in New York). That is an awful lot of rent coming from those hectares, and we see now why sites sometimes go for tens of millions of dollars. So right at the centre on our graphic model we can plot very high rents per hectare produced by the business offices and those small, but so fashionable and expensive, boutiques where Madame puts down $5,000 for a cocktail dress because she has 'absolutely nothing to wear this evening!'.

As a general rule, we do not usually find these sorts of expensive land uses farther out, so our office-boutique line on the graph declines with distance from the centre. But as we move out from the centre, we often find in a modern city the very expensive apartments and hotels. Generally, these cannot generate quite the rents of the offices and shops, so our apartment-hotel line starts lower down, but it also declines more slowly from the centre. At some point, say X on our graph,

26

the apartment-hotels can compete for the hectares with those expensive business offices and boutiques. Since they are the most profitable types of land use at this point, we would expect them to take over. But now something fascinating happens, and we have to think very geographically to understand it, and remember we are dealing with rent or money generated *per unit area*. What will be the next type of land use that can compete with the rich hotels and penthouses?

What do we often find in a modern city just off the downtown area? Usually the dilapidated, going-to-pot, formerly more expensive residential areas that today have become the landlord *slums*. If you do not believe me, just think of New York, Philadelphia, Chicago, Detroit . . . and so on. But, you say, this is ridiculous, how can slum apartments and broken-down housing controlled by landlords compete with the rich pent-houses and hotels? By definition, poor people can't pay as much rent as rich people! Ah, but wait a minute, says the geographer, we are talking about rent *per hectare*, remember? A few poor people paying rent will not be able to generate very much, but cram them in, do not bother to spend money fixing up the houses, bribe the health and building inspectors, and all those poor people can generate an awful lot of rent on that square hectare we are talking about. You do not believe me? Then think of Pittsburgh, St Louis, Milwaukee, Baltimore . . . where are the shameful slums and black ghettos of America? In the centre? No, they cannot compete with milady's boutique. Out in the ranch houses of the suburbs? No, that low-density residential housing is not for poor people. In almost every case, we find the slums just off the central business district, usually 'walled in' by the warehouses, freight yards, and truck sites on the other side (Figure 3.2). Only when we

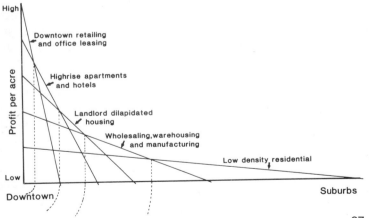

Figure 3.2: A slice through a typical American city, showing how rent per acre varies for different types of land use with distance from the centre.

break through this noisy and often dirty zone do we come to the well-kept residential roads, and eventually the green lawns, trees and perhaps the swimming pools of suburbia.

So here we have our highly abstract and simplified *model* of a city. Does it work? Do we really see rings of changing land use as we drive from the centre to the suburbs? No, not *exactly*, but try looking at and thinking about a modern American city through the 'lens' of this model, and by and large, we do indeed find these general regularities. Does the model account for all the cities of the world? No, certainly not. European cities, those that have developed over centuries, have somewhat different patterns of land use, and the exploding cities of Latin America and Africa, with their outskirts of wretched hovels constructed from anything the poor people can get their hands on, do not conform to the simple outlines of our rent model here. Even in the United States, things are changing rapidly in some of the urban areas. But as a simple model of land use it is not too bad, and it could be made much more complicated, and expressed mathematically, rather than in simple graphic form. Most geographic models today are expressed in mathematical form, and geographers try to deduce the consequences of what their equations say by manipulating them – often with the help of computers.

So this, very simply, is what a *model* means, and we can see now why the dramatic rise in that 'model' line on our graph (Figure 3.1, page 23) implies a *very* different way of approaching the task of geographic inquiry. It implies a process of abstraction, usually a high degree of abstraction, and often a manipulaton of mathematical equations. In fact, sometimes you wonder where all the real cities have gone.

The third word that exploded in a more roller-coaster fashion was the word *regional* (Figure 3.1), and underneath this rapid rise we find something fascinating going on. Remember that the notion of a region is a very old geographic idea, a 'natural area' within which all the physical and human elements are somehow interrelated and fused into a coherent whole. For many older geographers, regions were really the objects that geographers ought to study, in much the same way that chemists study molecules and compounds, zoologists their animals, and so on. But by the early 1950s, regional geography had deteriorated into the checklists and classifications of Professor Krick, and was 'on its way out'. In fact, by 1961 the word had almost disappeared altogether, although no one realized at the time that the pile of ashes contained the regional phoenix that would soon be born again. Even as the old region was disappearing in a great puff of intellectual banality, a new generation had rediscovered how terribly important it was to examine how 'things hang together' in geographic space.

It was really all a part of the growing awareness that we have to deal with very large and complex systems in our modern world, and that if you change one thing, or one connection, in a complicated and interrelated web of physical and human elements you may end up changing more than you bargained for. Today, the dangers of altering finely tuned systems hit us each time we think of the ways we have fouled up our local and worldwide environments. Illegal chemical dumping creates great anxiety that precious underground supplies of water may be contaminated for hundreds of years by highly dangerous, and extremely stable, compounds; the public water supplies of many towns on the Ohio and Mississippi rivers contain today scores of chemicals not found twenty years ago; acid rain destroys Canadian and Scandinavian forests and fish; the Rhine today is a poisonous chemical soup; fertilizers drain into rivers and lakes and deplete them of living oxygen; and the DDT sprayed up to ten years ago all over the world is still working its way up food chains from ocean water, to plankton, to fish, to birds and now to people.

So the 'regional explosion' was all part of the realization that making checklists was not good enough, that we had to know more – much more – about the connections and interrelationships, and know these as well as we possibly could so that if necessary we could recommend actions that would lead to a better state of affairs. And this is precisely the reason we find those other three words *structure*, *theory* and *planning* having modest explosions of their own (Figure 3.3), and notice the

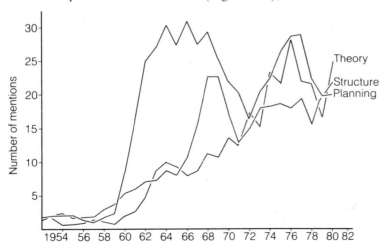

Figure 3.3: The frequency with which the words *structure*, *theory* and *planning* have been used in the titles of geographic articles over the past 30 years.

vertical scale is twice the previous one. If you start to emphasize connections and interrelationships between things in a regional setting, you are bound to start thinking in terms of *structure*, for what meaning does regional structure have except as it is expressed as connections between things? All sorts of connections: towns connected by roads, railways and airlines; countries connected by flows of trade; residential areas connected to work locations in the pulsating daily movements we call the journey-to-work; biological webs that mean chains and structures of cause and effect, and so on. No wonder geographers started talking about trade structures, urban structures, ecological structures, and many more.

With an increased willingness to tackle complicated systems and structures came the necessity to distance oneself a bit from all the complexity, and try to simplify (to model), abstract and *theorize*. You could not just plunge in and start recording information about all the natural and human complexity, for the first question was *what* are you going to record? Decisions that some things are more important than others are essentially *theoretical* decisions. The continuing rise in the concern for *theory* is again indicative of major changes, particularly as there is nothing so useful as good theory to guide you. It is here that we find the reason for the growth in our last tracer word (Figure 3.3): *planning* really does explode out of 1961, and the concern is maintained to the present day. One of the really major changes in geography has been the involvement of many modern geographers with practical, applied, essentially planning problems, *everywhere*. It is happening in cities and towns, as planners and consultants from everything ranging from new shopping centres to new forms of urban transportation; in underdeveloped regions, ranging from America's Appalachia and Italy's Mezzogiorno in the Western world, to rural areas of the Philippines and Tanzania in the Third World; in nations trying to plan the careful use of space and resources, and in international agencies, such as those of the United Nations and World Bank, trying to feed a hungry world, plan pure water supplies (and sewage disposal) for the exploding Third World cities, plan new agricultural regions in former areas of tropical swamp, and so on.

Let me make it quite personal so you will not think I am just spinning a tale. Former students of mine have directed the World Bank in Latin America, recommended rural road priorities in Indonesia and Malaysia, directed the first census for national planning in Afghanistan, helped Guyana with its emerging regional structure in new rice-growing areas, aided

Sierra Leone in its national planning efforts, worked on the water problems of new towns in Venezuela, on agricultural problems in Morocco and Sudan, on food marketing problems in Ghana, on the future requirements of the Paris Métro, on housing for desert migrants to Dakar, on the effects of maritime access for landlocked nations, and many more – it is a long list. One student of mine in the mid-1960s knocked on the door of one of the world's major international consulting firms with headquarters in Paris, the first geographer ever to do so. Speaking fluent English, French and FORTRAN (a computer language), he persuaded them that the modern geographic perspective was essential (in France only economists and engineers were normally hired), and he left them only twelve years later to found his own international consulting firm. The letters I receive from them around the world, some written by the light of a camp fire on a lonely plateau in Afghanistan, perhaps the first human presence there since the days of Genghis Khan, all attest to the importance of applied geographic insight.

The emphasis on applied geography was not entirely new, for land use geographers contributed to assessments of the Tennessee Valley Authority in the 1930s, and others inventoried land use in Britain during the Second World War when every scrap of land had to be put to use to feed a nation fighting for its life. Once in a while, and here and there, a geographer 'got involved', and sometimes made distinguished contributions to such things as microclimatology and the timing of crop harvesting by large commercial food producers, especially those producing frozen vegetables. During the Second World War, many geographers served in intelligence operations, for knowledge about physical and economic conditions became vital, and highly accurate maps were essential to the success of military operations. But the degree of involvement in applied planning problems was very small compared to today, and the geographer as consultant is really a new and still-growing phenomenon – and not one without its worrisome side effects, as we shall see (Chapter 16).

So words reflect our thinking, and those rising graphs tell us a story of great changes. So do the geographic journals and their contents. In the 1950s and 1960s we might have found such articles as 'Stingless beekeeping in western Mexico', 'Shell roads in Texas', 'The banana industry of western Samoa', 'Fences and farms', 'The future road system of Yellowstone National Park', 'Steatophygia of the Human Female in the Kalahari', and 'Quantification – a geographic deviation?' The conclusion to the last question was *yes*, and one wonders what sort of fieldwork was conducted in the Kalahari. In the 1970s and 1980s the journals now carry 'Remote sensing of water demand information', 'The dynamics of urban spatial structure', 'Residential location ambient air lead pollution and lead absorption in children', 'On spatial justice', 'The physical interpretation of eigenfunctions in dichotomous matrices', 'On modelling the evolution of hunter-gatherer decision-making systems', and 'The hexagon is alive and well and reproducing in the Sahara'! One senses immediately that things are not what they used to be. One or two of the older journals still publish on such things as 'Distribution of barn types in northwestern United States', 'The romantic face of Wales', 'The Turkish village coffeehouse as a Social Institution', 'Transhumance in Bosnia and Herzegovina', and 'Wind breaks in the lower Rhône valley', but one senses that these have a very different feel to them.

As for the journals themselves, they too have changed, expanded and grown in numbers, and such expansion is by no means peculiar to geography. In any intellectually vital field, new ideas, new questions and new ways of doing things produce tensions and pressures. The pressures are felt especially as the new perspectives try to find expression in the existing professional journals, most of which tend to be rather conservative and 'establishment', sometimes with editors who feel it is their sacred duty to serve as 'gatekeepers'. It is often like a head of water building up behind a dam: either the dam breaks, or new outlets have to be found to relieve the pressure. So it was that 14 new journals in English were born in these years, many started deliberately to bypass the old. A cartoon of the time commented upon the changing perspectives on the geography of harpoon handles (Figure 3.4). Of course, it was nonsensical jargon, parodying the new methods using large computers, mathematical models and highly formal logical and theoretical approaches, but it was also a delightful spoof on the rather constrained geographic tradition known as the Berkeley School – 'I mean, how was I to know he was from California?'

The rapid surges in nearly all the words we have looked at all start around the early 1960s, and the changes they reflect have come to be known as the Quantitative Revolution. It has long been over: the quantitative techniques and methodologies that were once so disturbing are now part of mainstream geography, taught to students as a natural part and parcel of the curriculum. At the time it seemed to many geographers, especially those who prided themselves on their qualitative assessments, that their lovely Geographia was being abducted

So the guy came to me with this study on the Geography of Polynesian Harpoon Handles.

. . . and the Bush-Mosteller Model III, modified by a delta junction choice maze held distinct possibilities for the quasi-communication problem at the micro-level.

In fact, flipping over the pages, it soon became obvious that a dynamic game incorporating the basic learning theory implicit in the first stages . . .

He threw himself off the lower deck of the George Washington bridge two hours later.

A Monte Carlo simulation was obviously the first step . . . provided NASA could spare time on STRETCH. . .

Of course, I told him I had an Arrow bias, but the formal set-theoretic statement stood out a mile, and would have clarified the whole problem enormously, as the fourteen axioms were child's play.

. . . would tie the whole thing up.

I mean, how was I to know he was from California?

Figure 3.4: The geography of harpoon handles, taken from the anonymous journal *Geography*, around 1962, a spoof on some of the traditional geographic approaches.

by a rude and Neanderthal Quantifactus (Figure 3.5). Splashing across the Fluvial Calculus to an arid and abstract 'model landscape', Geographia calls to her beautiful, but poor shepherd Qualifactus as he stands on the river bank verdant with meadow flowers lamenting her passing. Poor Geographia! Yet, as we shall see, she survived this and subsequent mishandling, and even got her own back.

Figure 3.5: The abduction of Geographia by Quantifactus, reprinted by permission of the anonymous artist who enjoys oxymorons.

Geography as a child of its technological time

4

Every once in a while it is not a bad thing if people, and the larger enterprises they form, stop for a moment, catch their breath, and ask themselves how they got to where they are, what they are doing, why they are doing it, and if they really want to go on in the same way. These are not easy questions to answer, partly because we are so caught up in our work that we find it difficult to distance ourselves a bit, to get some perspective, and to think carefully about where we are going. Archimedes claimed that with the power the lever gave to his muscles he could move the world – *provided* he could find a place to stand on. That, of course, is always the trick. As individuals thinking about ourselves (an essential task if we are to grow and enrich our lives), or as geographers thinking about geography (an equally essential task if geographic inquiry is to enrich our understanding), how do we find that Archimedean point 'on the outside' from which to get some perspective and take stock of what we are doing? We shall see later (Chapter 26) that careful and critical reflection about geography and geographic research is a growing preoccupation among thoughtful geographers – *thoughtful*, almost by definition, because they are trying to think about thinking. Yet even at this point in the book it is worth stepping back for a moment to ask some questions about those verbal explosions we have just looked at, and wonder out loud why phrases such as the 'Quantitative Revolution' and the 'New Geography' suddenly appeared when they did in the late 1950s and early 1960s.

It seems to me that a more basic and general question is why does a particular field of inquiry like geography study things in certain ways and not others at a particular historical time? This question really applies to all fields of human endeavour, right across the broad spectrum of the sciences and the arts. For example, as an astronomer you do not look at the stars in a framework of relativity and quantum physics if the year is 1870. Why not? Well, because Albert Einstein and Max Planck have not come along yet. But, you might say, suppose they *hadn't* come along when they did, would we still be back in the age of classical Newtonian mechanics, a framework so powerful that it stood for 300 years, and led to results so all-embracing that physicists towards the end of the nineteenth century were thinking of 'just a few more mopping up operations, and we'll have it all sewn up'? The answer almost certainly is *no*: on hindsight we can see that small cracks were already appearing in the fabric of classical physics, new thoughts were 'in the air', and if Einstein had not made his thought-shaking proposals in the early twentieth century, you can be almost sure that someone, somewhere, would have made the same suggestions shortly thereafter.

It is exactly the same in the arts: a restlessness often appears in many areas simultaneously, although these dislocations and perturbations are often connected together in strange, hard-to-define, but intuitively valid ways. In fact, we know very little about these connections that make up the multidimensional worlds and structures that we call cultures, although Vienna at the turn of the century provides us with an almost classical case study of the deep relationships between ideas in philosophy, mathematics, physics, music, theatre, literature, journalism, poetry, architecture, painting and the growing field of psychiatry (Freud). New ways of expressing the artistic impulse of a people seem to appear 'when their time is ripe', but they *always* stand in an historical tradition, and ultimately spring from it. Should you object that this is ultimately a tautology – things happen when things are ready to happen – my only answer to you is, so be it. 'There is a tide in the affairs of men . . .', William Shakespeare knew it in his bones, and if you do not try to resist the idea too much, your bones tell you the same thing.

So it was with the sudden appearance of new ideas, new approaches and new perspectives in geography. The time was ripe for them. The different ways of thinking about patterns of human settlement provided by Morisita in Japan, Walter Christaller and August Lösch in Germany, Edgar Kant in Estonia, and Robert Dickinson and Arthur Smailes in England during the 1930s, produced somewhat delayed reactions because of the Second World War, but afterwards the times 'continued to ripen'. Today, a highly mathematical theory of human settlements, called central place theory (Chapter 9), forms a major building block of contemporary geography.

After the war, the geographer Reino Ajo supported his family in Finland as an inspector of automobiles (he could not

Robert Dickinson
University of Arizona
1905–1981

36

get a permanent faculty position in the traditional departments of geography in his country), but he continued his research in the evenings, employing mathematical methods not seen before in the field, publishing in five languages (with abstracts in Latin) in an effort to reach a larger world. In Sweden, Sven Godlund and Torsten Hägerstrand started to wonder about the dynamics of geographic patterns – why and how settlements grew and collapsed, how things and ideas spread over time and over geographic space – and when the latter was invited to the United States at the end of the 1950s, he found one of those rare and fruitful juxtapositions of people that seem to happen from time to time just by chance. William Garrison, deeply influenced by Walter Isard (who had just founded the first department of regional science in the world), attracted an extraordinary group of students in the late 1950s, the 'space cadets' whose names continue to be prominent today. We shall meet a number of them in the pages ahead. At other places, similar groups appeared quite spontaneously and independently, made up of geographers with the same sense of restlessness, the same sense of 'something's got to change'.

Arthur Smailes
University of London
1911–1984

So, you say to me, haven't you really contradicted yourself? First you talk about 'when the times are ripe', and then you end up trotting out the names of specific people who seem to have started the whole thing off. Which way are you going to go – things happen with a mind of their own, or history is the result of a few remarkable individuals? And my answer is you really cannot have one without the other. No individuals, and nothing happens. No movement and change, no historical readiness to accept, and again nothing happens. All the individuals concerned appeared at particular times and places, and formed particular associations and connections, but each in their different ways stood at the edge of particular historical developments in their field, were influenced by these, and helped to contribute to them. Historical change is not just a simple matter of one or the other.

Reino Ajo
Independent Scholar
1902–1974

I think we must remember that geography and geographers are not isolated, for they all stand within the larger 'immediate thinking life' of their times and the encompassing societies of which they are a part. If you think about it for a minute, what is one of the outstanding characteristics of modern life but *technology*? We are all thrown into a *technical world*, a world not of our own making, for how can *we* be responsible for genetic engineering, atomic bombs, lasers, computers, etc.? They were already here when we got here, or arrived without our knowledge, and we can do virtually nothing about them,

William Garrison
*University of California,
Berkeley*
1924–

Walter Isard
Cornell University
1919–

in the sense of getting rid of them, because today they, literally, *are*. So it is hardly surprising that the major and recent changes in geography – in geographic research and in geographic teaching – are essentially technical changes, the result of the technical world in which geographers find themselves. Once you see geography as something caught up in its own times, they could hardly be anything else.

Please do not misunderstand me here: I am not proposing that we embark upon a mindless tirade against technology, or engage in the sort of weepy hand-wringing that some employ as a form of catharsis that usually stops real thinking dead in its tracks. Rather, let us take some time to think together about some of the characteristics of the world in which geography and geographers find themselves today. First of all, it is a world in which language is becoming more and more impoverished, as television displays an abominable sense of grammar, vocabulary and syntax (ask people to define *syntax* and most will shrug their shoulders, or do a 'well . . . er, you know'). When children, on the average, spend as much or more time in front of a television set as they spend in school, then the rich fabric of language is bound to thin and pull apart, particularly as official Newspeak so often provides a soothing cloud of ambiguity, euphemism and misrepresentation that moral concern is wiped away. Does this help us understand why so much discourse in the human sciences and the arts (and literary and art criticism are especially obfuscating) appears as jargon? Not the crisp, clear, immediately understood jargon of the professional sailor up aloft receiving instructions in a howling gale from the bo'sun below – there are times when jargon is essential as an emergency shorthand – but the pretentious, shallow and ultimately dishonest jargon that puffs up platitudes into thoughts as solid as the facades of those western towns in Hollywood movies.

This technological world is also a world of TV games and electronic amusement arcades, where children (some in their sixties) plug themselves in with an intellectual umbilical cord and become part of the machines themselves. At that point, who controls whom? Even the Surgeon-General of the United States has issued 'health warnings' about the insidious effects of these machines on children. We wonder why children are more aggressive these days, but after tens of thousands of acts of violence have been viewed by children as ways to 'resolve problems', why are we surprised if teenagers act violently to resolve problems? Today, in the United States (and I am sure elsewhere around the world soon), video games are sold to 'adults' which involve a white cavalry officer fighting his way

DANGER !!
Television games may be injurious to your health
Signed:
The Surgeon General
of the United States

38

through hordes of brown Indians in order to rape an Indian girl tied to a stake. If the human world reflects the technical, the technical world reflects the human.

It is the same world in which the computer is seen by many as the saviour of education (just as the new technology of television was seen as the saviour twenty years ago), so that children, whose attention spans have been conditioned by TV commercials, are taught to press keys to respond to feedback – just like white rats receiving their food pellets for 'desirable behaviour'. It is not so much a matter of artificial intelligence modelling people, but how quickly we can transform people into machine-like intelligences. An appalling trend in art education today embraces computer 'art' as a way of educating children, the medium through which they are given opportunities to release their own expressions of colour and form, as well as introducing them to their own artistic heritage. The paints, the brushes, the different papers with their smells and textures, the modelling clay, the varying hardnesses of the pencils, and the different effects the individual child achieves – all gone, all reduced to punching a key, gazing at a cathode-ray tube, and twiddling a joy-stick (an ironic misnomer) to move an automatic pen on a plotting table. What sort of world are we handing on to our children if these sorts of mindless fools get their way?

Again, it is the same world in which computer programs can be written to imitate the responses of a psychiatrist, so that people start to pour their hearts out to the purely mechanical impulses of an electronic machine. Initially written by Joseph Weizenbaum, a 'grandfather' of computer technology at the Massachusetts Institute of Technology, the program ELIZA responds from a limited list of phrases into which words used by the person punching away at the keyboard are injected and recycled after a delay of four or five responses. 'Conversations' seem so realistic, and ELIZA seems so understanding, that people start to reveal some of their most intimate and private thoughts – all carefully recorded on an unseen printout for the nice man in the next room. Such responses came as a great shock and surprise to Weizenbaum (and I do not think any of us would have predicted them either), but what really appalled him was the reaction of a number of psychiatrists. They thought it was a marvellous idea! Just think, we can plug everyone with problems into a time-sharing computer – hundreds at a time if necessary – and let them relieve their anxieties with the kind and patient machine. To be fair, many other psychiatrists, those working at levels of deeper and more humane concern, were

Joseph Weizenbaum
Massachusetts Institute of Technology
1923–

as appalled as Weizenbaum, but it was this sort of man-machine experience that led him to reflect deeply on some of the wider implications of computer technology for people – who are *not* machines.

This is not to say that computers *as* machines cannot play an enormously helpful, deeply grounded and humane role in other areas of medicine. And here we come face to face with the tension that is almost always produced by technological advance. ELIZA-like programs have been used with careful concern to help and monitor the progress of people whose brains have suffered the physical damage of a stroke, an accident, or perhaps tumour-excising surgery. Moreover, the ever-watchful computer can monitor dozens of life-signals of dozens of people in post-operative wards, and sound the alarm immediately when something is wrong. No tired and overworked nurse, no matter how conscientious, could keep a never-sleeping eye on so much literally *vital* information.

But this same technological world works its way into so many aspects of our modern lives, for both good and ill, that it seems to take on a life and destiny of its own, shaping and limiting the choices we can make. If you are thrown into a highly technical world of the late twentieth century, and you grow up with TV, home computers, automobiles, automatic this and automatic that, and skies and sunsets never free of jet contrails, how do other choices occur to you – you, the child shaped by this technological world? Today, once calm and tranquil lakes and seas are torn by the screaming motors of jetskis and high-speed outboards pulling water-skiers. The hush of deep forests in winter, where once the silence was broken only by the gentle hiss of skis, is now shattered by the sound of snowmobiles leaving behind their blue exhausts of oil-stench to smother the fresh smell of pine and hemlock. In summer, the trail bikes squeeze out the remaining pockets of wilderness where men and women could once escape. And then we have the gall to call such man-machine symbiosis *re-creation*.

How do we generate the sorts of values that led countries like Norway and France to ban snowmobiles – except for emergency rescue work in winter? In Norway, the quiet soul-restoring peace of the mountains and forests belongs to everyone, and the values of the many declare that their heritage of tranquillity will not be lost to the technological age shaped by a few. I often think geography students could help us see the problem at the local level by compiling a series of maps for the local newspaper or TV station showing how the slobbism of technology has diffused out of the towns into the surrounding countryside. Maps, for example, showing the cancer-like spread of automobile graveyards, junked machines, plastic rubbish, noise and smell – all of it working its way along the roadsides, creeping into the fields, oozing over hills and forests and along hiking trails and oil-ridden beaches. No wonder Danes and Swedes forbid the purchase of summer homes by rich industrialists from the rest of Europe, where the ethic of *Håll Naturen Ren* – Keep Nature Clean – has long been pushed into concealment as the children follow the uncaring example of their parents. I sometimes think the paths of Portugal, Spain, France and Italy are going to disappear under the plastic rubbish that uncaring and insensitive men, women and children leave behind.

All right, all right, Dr Pangloss, you say to me playing your role of Candide, despite all those assurances about not embarking upon tirades it seems to me that you're getting pretty close. What's all this got to do with geography, and isn't it time we got back to the subject? Isn't it time we went out to dig the garden?

Well, you are right in a way, but taking this quick look at some of the worst effects of the technological world we live in has a purpose, because this is the world of technique, machine and mechanism in which geography and geographers exist today. This is our *milieu*, our *Weltanschauung*, and a collective human and intellectual enterprise like geography cannot help being a part, in a sense a child, of it. If we are to understand some important characteristics of modern geography, and how it came to be that way, we have got to try to find that Archimedean point to see it in its larger setting. That setting is essentially one of machine and mechanism, of technique and manipulation – for good *and* for ill. Or so it seems to me.

What are some of the consequences? First of all, a 'quantitative revolution' obviously uses mathematics. Well and good – this, as we have seen, is an old and honourable tradition in geography going back to its Greek and Arab roots, a tradition nurtured by the need for accurate navigation and map-making from the Renaissance up to the earth satellites of the present day. You cannot construct maps if you do not know spherical trigonometry, and it helps even more if you can grasp a bit more deeply what we might loosely call 'transformational mathematics' – how shapes like round globes are transformed into flat and more useful maps. But at the same time, you had better be a bit careful about the sorts of mathematics you use, because a lot of it until quite recently (let us say the last hundred years), was developed to describe the physical world

41

of things, especially the worlds of celestial and statistical mechanics from the eighteenth and nineteenth centuries respectively (Chapter 25).

One of the problems that geography faces, along with all the other human sciences, is that its mathematics is borrowed, and much of it was originally generated by the need to describe a physical world of *mechanism*. This means that if geographers borrow what is essentially a mathematics of mechanism to describe certain aspects of the human world, and that mathematics comes straight out of mechanics – levers, forces, attracting masses, atoms like billiard balls and so on – then the human world expressed in this borrowed 'language' *cannot look like anything except a big machine*. And since language shapes thinking, geographers employing such mathematical 'languages' are going to have their thinking channelled and directed towards mechanistic models. So in a sense, within this pre-chosen, but unthought-about mechanistic framework, the thinking of geographers may already be trapped, pre-structured and disposed towards a mechanical view of human society. The same holds true of nearly all the social sciences: many economists think the economy works like a big machine, so that if I turn cog M (money supply), eventually the cog U (unemployment), all mechanically coupled down the line, will also turn in the right direction. No amount of empirical evidence to the contrary will persuade them otherwise. Much of psychology is in the same state, and so is sociology. Few can see the trap they are in, and all have paid a terrible price for severing their traditional ties to philosophy.

Again, not a tirade, but an attempt to think what the consequences might be. It is obvious that when we try to describe something as complex as changes over time and geographic space, we have to take a more general, and that means in some degree *simplified*, view. But we should also do our best to be schizophrenic: to work sometimes with mechanical forms of mathematics, yet be constantly aware of what we are doing 'down there' from the higher perspective given to us by our Archimedean point of self-reflection. For mechanics, quite properly and legitimately, is a science of knowing and manipulating physical things, and there is no question that our modern world could not exist without our capacity to manipulate by devising technical solutions to some problems. The difficulties come when thinking about technical solutions shifts sideways into the parallel human world where the 'things' are not things at all, but you and me, human beings with consciousness, with the capacity for self-reflection, and the ability to judge and

42

make choices on moral, ethical, aesthetic, religious and many other grounds – including those of love and concern.

If, in the course of our investigations, we simplify and generalize – and it seems to me that we must, because we really do not have the capacity to grasp all the details in any human situation – then we have to be constantly aware and sensitive to what we are doing. If we are not, then all sorts of detrimental and thoughtless consequences may follow. For example, the manipulation of human society by technical solutions may well mean that the geographic *status quo* is incarcerated in the spatial patterns, arrangements and divisions that exist now, so that these approaches may prevent us from thinking about a better human world that might be in the future. Some geographers in Sweden have been deeply critical of the way the old political subdivisions (some of them dating back to medieval times) have been technically 'rationalized', claiming that the new divisions, devised on the basis of impeccable data and technique, freeze the geographic *status quo* of their society. In France, it is remarkable how tenaciously people cling to the old *commune* system, resisting plans to aggregate these old units into the rational order of the technocrats from Paris. Writing this book in one, very beautiful commune in the mountains of the Vercors in Southeastern France, I understand, at a very deep and personal level, why people do not want to be amalgamated and 'rationalized'. One geographic study pointing to the aggregation of power with the aggregation of old political units in France was so illuminating that it was deemed desirable to suppress it. Which the technocrats in the Ministry in Paris promptly did.

A final danger – and once again I want to emphasize that I am focusing on the dangers here, and ignoring the considerable benefits – is that when geographic education takes place in a surrounding matrix of technological thought, then education may become increasingly *training*. We all know the difference instinctively: *education* liberates thought and encourages thinking to direct itself to new and deeper issues. *Training*, on the contrary, tends to impose the 'right way' of doing things, the accepted, the I-did-it-this-way-so-will-you (and if you do not, you will fail). Sit back and think for a moment what our schools and universities do most of the time – educate or train? This issue is *not* simple, and it is not just a matter of choosing one and discarding the other. A certain amount of training, a certain acquisition of technique, is essential in *any* field. Effective – that is to say skilled, competent, professional – technique is crucial for investigations in biology, wildlife management,

43

physics, urban planning . . . in *any* scientific field, and if you are undergoing by-pass surgery, would you not prefer your surgeon to have good technique? Similarly, the artist develops by long and careful training, repetition, and constant use, such a deep technical facility with the materials used that the technical aspects are eventually subsumed and used simply because they are 'at hand' rather than constantly thought about. No musician would disparage technique – to do so would be to deny the possibility of creative musical performance.

So it is in geography. In the universities particularly we have to help our students acquire the whole range of analytical, mathematical, graphic and computer techniques now available. Without these, the range of investigative possibility is severely constrained, and *education*, we said, was a releasing of constraints, not a tightening of the screws. Yet I must admit there are times when I see roomfuls of students punching away at computer terminals and my heart sinks. Some, a few, are using the computer to open up an area of exciting investigation that would simply be impossible without these machines. But I suspect the majority are hooked, purely and simply hooked, on the wretched things, sitting there hour after hour, fiddling away, 'playing', and thinking that they are doing something creative and worthwhile. And it is right here, as you and I sit together at a computer terminal (the keyboard that allows us to interact with the machine), it is at this point that we come face to face with the tension, paradox and contradiction – call it what you will. On the one hand, the computer can seduce and entice, creating a new priesthood with their own jargon that is both useful shorthand and a barrier to keep out the uninitiated. It has the capacity to reduce the human being to its own mechanical level. On the other hand, it is the thing that lies embedded at the heart of the explosion in modern geography, and this book could not possibly have been written without it. Almost everything we are going to look at is based upon our new-found ability to use the computer as a prosthetic device, to release our thinking towards new possibilities.

It is that tension-ridden, contradictory, what some would call *dialectical* capacity for entrapping our thought and releasing our thought that characterizes the modern computer, and we had better take a firm grip on that Archimedean Point, on our capacity as human beings to reflect upon what we are doing, as we approach and use it. The computer is at the heart of modern geography today, because geography is a child of its technological time – both for ill (the side effects are appalling), and for good (the benefits are enormous). It is time we looked at the computer and geography a bit more carefully.

5 What the computer did

In this age of the home computer, we tend to forget how very recent the whole idea of a computing machine really is. Of course, some aids to doing arithmetic are very old, and simple forms of the abacus probably go back thousands of years. To watch a skilled abacian (?) calculate with lightning-fast flicks of the fingers on a modern abacus is to marvel at a finely honed human capacity, although I must say that my little Texas Instruments TI-55 that I use constantly to keep my cheque-book straight seems equally miraculous. How does it do those calculations so quickly? But the real origins of the computing machine seem to lie with two great mathematicians, and the way their work was extended by two equally remarkable Victorians. The French mathematician Blaise Pascal invented a simple device of gear wheels to carry the tens over to the next place when doing addition, while the German mathematician Gottfried Leibniz extended this mechanical procedure to subtraction, multiplication and division. He also saw that for the purposes of calculating it was easier to do this sort of counting if you use a number or counting system based on 2, rather than the one we normally use based on 10 – presumably because we have ten fingers to count on. If you have not counted in binary before, let us do a few together, because this invention of 1671 is still the basis of all modern computers. It starts in the familiar way:

binary	'ten fingers'
(0, 1)	(0, 1, 2, 3, 4, 5, 6, 7, 8, 9)
0	0
1	1
.	.
.	.
.	.

But then in the binary system we have used up all our 'numbers' (0, 1), so we have to carry over to the next place like this:

10	2
11	3
.	.
.	.
.	.

At which point we are stuck again, unless we carry over to the third place:

100	4
101	5
110	6
.	.
.	.
.	.

and so on. Although it seems very awkward the first time you come across it (do we really have to call 1001 'nine'?), I think you can see the advantage of this (0, 1) binary system when you think of electrical switches being either off (0) or on (1), and realize that modern computers are little more than lots of off-on switches all linked together.

The idea of a calculating machine was extended by Charles Babbage and his formidable mathematical friend Ada Lovelace in the middle of the nineteenth century. His 'analytical engine' and 'difference machine' were never actually completed in full, because the money ran out, and the Admiralty (interested in the trajectories of shells!) refused to support it further. Nevertheless, they had the idea of storing information on cards and wheels, because coding information on punched cards was already known from making complicated tapestries, for which the looms were guided by cards with holes cut into them. So nearly all the bits and pieces for making the modern computer were in place by the last years of the nineteenth century, and all that was needed was the speed of electricity and an urgent task that would justify spending lots of money. The electronic vacuum tube provided the first, and the Manhattan Project to make the atomic bomb during the Second World War provided the second. The original electronic computers were built to do the millions of calculations required.

By our standards today, the computers of 1945 were pathetically slow, capable of 2–3,000 binary operations like addition and subtraction per second, compared to 620 million per second today on the specially built Hitachi computers. I do not know about you, but I just cannot imagine doing 620 million additions

Ada Lovelace
Independent Scholar
1815–1852

Charles Babbage
University of Cambridge
1791–1871

47

per second! Yet even faster computers are now on the drawing boards, and in 1992 the ETA will carry out 30 *billion* operations per second. But in the 'old days', in the late 1950s, a few universities were acquiring the IBM 650, a massive arrangement of vacuum tubes that put out so much heat that air conditioning on a large scale was required. I can remember in 1960 calculating some special values (called eigenvalues) from a square table of numbers on a manual desk calculator, and it took me nearly two months of agonizingly careful work, terrified at every step on the way that I might make a mistake. These values were used as a check on the computer, which did the same task in eight *minutes*, and the values checked reasonably well. Today it would take one-tenth of a second on most university computers, and millionths of a second on the Hitachi. Years later, when bigger machines were commonplace in most universities, two graduate students at the University of Iowa published a correction: some of my hand and computer calculated values were off in the second decimal place. If I could have sent a thoroughly vulgar gesture not normally employed by gentlemen to Iowa I would have done so.

Today things are very different. Most geography departments have their own computer terminals connected to a time-sharing, mainframe computer which handles hundreds of people at the same time, and these basic facilities are augmented by graphic plotters for making maps and diagrams. Courses in computer programming – the actual writing of the instructions to tell a computer what to do – are standard parts of geographic curricula today, and most students go on to take more specialized work necessary for courses in analytical methods, remote sensing and computer cartography. We shall look at some of these later. Today, the micro-computer, a small computer that sits on a desk top, but with the same capacity as quite big computers only a few years ago, is appearing everywhere, taking the pressure off the big mainframe machines which are used increasingly for the really massive tasks of computing.

The impact of the computer in geography has been basically in two areas: teaching and research. One of the marvellous benefits of the computer for the teacher is that geographic problems and exercises can be made so much more realistic, and it also means that a lot of seventeenth-century mathematics of great analytical difficulty can simply be by-passed. In physics, for example, one teacher found himself faced with teaching an elementary physics course as part of a general science requirement in an American university, and discovered that most of the students had little or no calculus. To the traditional physicist, the idea of teaching university level physics without a background in the differential and integral calculus would have been absurd, a contradiction in terms. After all, Isaac Newton *invented* the calculus to do physics! But the students were able to do arithmetic, and computers are very good at doing arithmetic very quickly. So the teacher translated all the traditional expressions and equations into arithmetic, got the students going on the terminals, and used the computer to do all the little bits of adding and subtracting. Intrigued by the increased understanding he generated for himself and his students, he continued into relativity theory and pushed on to extremely difficult areas of hydrodynamics – the behaviour of flowing fluids. In this last area, the mathematics is often absolutely horrendous, with analytical expressions that tax the powers of first-rate mathematicians. In contrast, the computer not only chomps them up with its finite arithmetic, but prints out all sorts of nice graphics showing all the eddies and vortices forming in the fluids as they flow around different sorts of shapes.

It is the same story all over geography today. In a beginning class of mine, a series of lectures given as an elective or social science requirement in a large American university, I build the course around five computer exercises. All of these employ programs that were on the research frontiers, or did not even exist, twenty years ago. In the first exercise, students make their own, often highly imaginative maps – have you ever seen a map of toppings in pizza space, or a sonatina in Corelli space? – and one freshman used the exercise to make a map of the subjects of his own college, based on the prerequisite courses the various majors shared, to help him find his way about (Figure 5.1). The map told him that if he started in computer science, and later decided he wanted to be a doctor, it was going to be a long journey in prerequisite space. So, says the map, if you like science, but you're not exactly sure which way you want to go at the beginning, take the general science option. It's right there in the middle, and so forms the most *accessible* choice to the more specialized fields you may choose later. The student followed the map's advice.

Other computer-based exercises include locating things to serve people in a region (child care centres, schools, hospitals, etc.); how to make the best decisions when you are constrained by what you have available – a common experience of us all; modelling how some idea or innovation spreads through a population, with a computer making maps for each time period; and 'playing' with a complicated model of the world that takes into account such things as population growth, investment,

49

Figure 5.1: A map of science 'majors' in prerequisite space constructed by a student to help him think through the course of study he wished to undertake during his first two years.

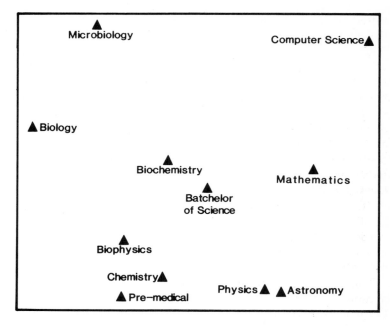

pollution and agricultural production. All the exercises are made up, or real examples devised, by the students themselves, and they offer opportunities for their creative imagination and concern. It would literally have been impossible to offer them such opportunities before the advent of the modern computer.

The second area where the computer has made an enormous impact is in geographic research, and here we are entering an area of great intellectual and practical importance. I think it is fair to say that most of the questions now being asked in geographic research could not have been asked before the computer came along. If that sounds strange, remember we often do not ask questions if we have not had the grounds of possibility enlarged. You do not *think* about many things until it has been pointed out to you that such thoughts, and the accompanying questions, are real possibilities. I think the possibilities opened up are really of two kinds: a greatly enlarged capacity to handle information, and an equally heightened capacity to calculate with it. Let us look at these two aspects in turn, realizing that in actual practice they usually go together.

Information: this has always been a great problem for the geographer, because of that feeling of wanting to understand how lots of different things 'hang together' in geographic space. As we saw, it is a tradition going back to the questions of relations between soils, climates, altitude, vegetation and so on raised by von Humboldt, questions that still underpin much of the modern concern for complex systems of man-environment relations. Whatever the tradition, we often have to handle lots of information at the same time, and that means being able to store it in a readily 'get-attable' way. The tapes and discs of the modern computer are ideal for storing such information in such a way that it can be recalled almost instantly when needed. Such enormous storage capacities are at the heart of all modern cartography (Chapter 17), for even the most ordinary road map contains millions of bits of information, and most world maps today are outlined by drawing upon 10,000 very precisely located coastline points stored on a magnetic tape or disc. Furthermore, remote sensing, the use of imagery from earth satellites (Chapter 18), would be unthinkable without the information storage capacity of computers. Even now we tend to be overwhelmed by data, and the business of just storing and retrieving it has become quite an art in itself.

The other side of this is what do you do with all the information? It is here that the enormous computational power of the computer comes into its own. In a strange sort of way, computer computation today has almost returned the geographer to the Age of Exploration – not the sort of exploration using dog sleds on the icecap, or surveying the Hindu Kush, but exploring the structure of huge sets of data to find meaningful relationships within them. If you suspect that some things are meaningfully related to other things in a great mass of complexity you are trying to observe and understand, you have to have ways of exploring, testing and trying out your hunches. There was a time when geography went through a terribly pure stage of the ideal statistician, when the poor students were told to dress up their questions and hunches in high-falutin language and 'pose hypotheses', on the grounds that this was what 'real science' was about. But gradually geographers grew up, pushed the pure statistician to the side until he was needed, or taught the statistician that much of his paraphernalia was useless for *real* science. For real science is actually a messy business of speculation, wondering, following gut feelings, going up blind alleys – in a word *exploring*. This sort of messy exploration – and it must be messy and uncertain because otherwise you would know where you are going and would not need to explore at all – this sort of messy exploration into lots of information and large data sets needs a computer as a – I am tempted to use Charles Babbage's term – an analytical engine.

Let us take an example involving the problem of laying out

and examining many possibilities, and then trying to choose one that is acceptable to people. The problem is a very practical one, and one that could hardly be undertaken without a computer to help us. We shall call using a computer in this way *combinatorial search*, because we have to lay out and examine all sorts of possibilities that arise from the different ways we can combine things. Not too long ago, the School Board of Detroit, dominated by white members in a racially mixed city, wanted to 'redistrict' the school areas. Ostensibly it was to *rationalize* (a lovely covering up word, often used when something unpleasant is being forced on people) the school districts, but as soon as the plan was announced it was obvious that all the new school districts had been redrawn in such a way that the white people had majority control everywhere (Figure 5.2).

Figure 5.2: The Detroit School Board's rationalization and consolidation plan giving majority control everywhere to white residents. Note the discontinuity across the park to achieve dominance.

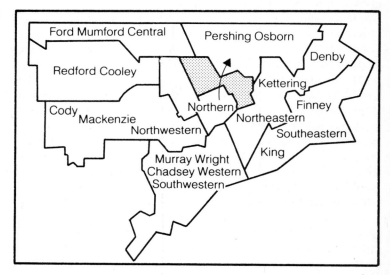

In fact, the proposal was an old-fashioned piece of gerrymandering – the careful subdivision of geographic space to ensure a political party or a particular group of people are in the majority. In Detroit, some of the recombinations were tortuous, for example the joining of Northern to Pershing and Osborn across a separating park to make sure white control was secure.

Now while the real intent was plain, it was a bit more difficult to respond to it with alternative suggestions, particularly in a court of law, where the whole mess eventually wound up. The question is how *do* you respond, how *do* you recombine districts and come up with a plan that is equitable? You can see at once that the division of geographic space is often loaded with moral

52

questions, questions of what is fair and right and just. First of all you need information – how many black and white families live in each of the existing districts that are going to be shuffled around and joined into new ones. Then you need to know how many ways you can combine the districts into new ones – an easy task for a mathematician *except* for one thing. Even the Detroit School Board realized that whatever new districts they devised, the districts had to be all in one piece. A geographer would say you had to find all the combinations under a spatial contiguity constraint – all the small pieces making up a new district had to be geographically contiguous or touching each other. That made the problem a particularly tough one to solve.

So two geographers programmed a computer, gave it the districts, the black/white ratios and the contiguity information, and told it to go to it. A total of 7,011 different redivisions came out the other end, each with the proportion of black and white citizens listed. Armed with this information, the black community of Detroit were able to argue in court for a much fairer and equitable solution.

So the computer is with us, for good and for ill, and when carefully used for scientific, practical and teaching purposes, it must be seen as a blessing. Not carefully used, it can become a curse, not the least of which is the danger that it will make geographers look at the world they are trying to understand entirely through the lens of computation. There is a real danger that by looking to the 'big calculation' for the solution, a simpler, more appropriate, and more obvious solution lying close at hand escapes us. We can see this in the famous four-colour map problem, which says you only need a minimum of four colours to fill in the regions of a map so that no two adjacent regions have the same colour (except at a point), no matter how tortuously you divide the map up (Figure 5.3). A formal proof has evaded mathematicians for centuries, but was finally achieved by an enormous amount of computer bludgeoning. Many mathematicians felt that somehow they had been cheated: mathematical proofs ought to have a quality of beauty and elegance about them, and enormous piles of computer output, that no one is going to check through anyway, are about as inelegant and grubby as you can get.

Now it happens that a number of geographers, including myself, know of a short, geometric, and extremely elegant proof of the four-colour theorem originally discovered by Professor Herman Krick. But our lips are sealed forever, of course. Mathematicians are sensitive souls, and do not like being shown up, and anyway we do not want to spoil their fun.

Figure 5.3: No matter how a map is divided, four colours are always sufficient to distinguish the regional subdivisions, allowing for the fact that four regions are allowed to meet at a point.

A pity really . . . it's so beautifully obvious – once you see it!

The final danger is to succumb to the temptation that the great wad of computer output lying on the table must mean *something*, when really all we have is a bad case of GIGO – Garbage In, Garbage Out. In the early years of computer facilities, many geographers fell into this trap – I certainly did – but most are climbing out today, being much more careful about the first steps before the computer is ever brought in. We all live and learn. And one of the things that has helped us learn, and so guide our inquiries more carefully, is the development of bodies of geographic theory. It is time we had a look at what these might entail.

Part II
A concern for theory

Theory in geography: a matter of some gravity

6

Why is it that geographers keep on insisting that their precious spatial perspective on the world is so important? Everyone else seems to ignore it, and gets on perfectly well without it, so why this constant nag, nag, nag about geographic space? How does it affect us, and why should we take it into account? After all, we all seem to be here in time and space, so what is all the fuss about?

It seems to me that these are perfectly legitimate questions. If the geographic dimensions of people's lives are really important, then at some point the geographer has got to stop jumping up and down about this thing called geographic space, and explain why we should start thinking about it more than we do. Sometimes things that we take for granted, things that are so much a part of our lives that they seem naively obvious, are actually very important. So let us start with our own direct experiences and see where they lead.

Suppose I asked you if you would mind keeping a rather detailed diary for a month or so, recording such things as where you go to buy groceries, have a beer at the pub, see a movie, and visit Aunt Eliza, and also ask you to write down all your telephone calls – who you called, and where the people were. At the end of the month, we could sit down together with a large map of your area, locate your home, and plot all the journeys and visits and telephone calls you have made as small black dots. What would the resulting map look like? Almost certainly it would more or less resemble a nebula-like cluster, dense in the middle around your home – lots of chatting with your neighbours, borrowing the occasional cups of sugar when you ran out, telephoning down the street that little Johnny has fallen off his tricycle – and then thinning out at the edges as you make a call to a shop in the next town, visit a friend you have not seen in a while, and so on. Obviously, the map we make of the way you visit and talk to people (the way you *interact* with others), would not be a perfectly round and smooth nebula of points. The way you interact with others depends on where they are located, your own personal preferences, and, in a rather deep sense, the way geographic space is structured from your point of view. Nevertheless, people generally tend to make lots of short visits and calls, and only a few long ones.

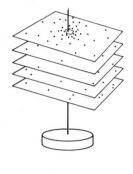

Now if we had asked lots of people to keep these same sort of personal diaries, we could construct a map for each one of them, and then 'add them up' by centring all their homes at one point – rather like skewering all the home locations so they coincided with each other. With all these centred and lying on top of each other, we could then compile a master map of all the calls and visits, and this would certainly show very strongly the nebula-like pattern I mentioned before. I want you to imagine now that this master map of ours is actually like an old-fashioned folding fan. We could cut the map from the edge to the centre (where all the home locations are) and 'fold it up'. What we are doing, of course, is transforming our two-dimensional world into a one-dimensional one (Figure 6.1). As

Figure 6.1: Folding the two-dimensional nebula-like cluster on the master map into a one-dimensional distribution of frequencies of contacts with distance away from home.

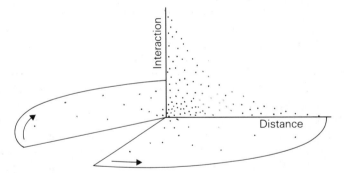

a result, all the little dots representing individual interactions pile up in a distribution showing the frequency of the interactions related to the distances over which they take place. What we find, again and again, all over the world, is what geographers call the 'distance decay effect' – the greater the distance, the less the number of interactions.

But, you say to me, supposing we had the diaries of some of the jet set, perhaps those 'scholars', committee-lovers, and international hangers-on that Arthur Koestler called the Call Girls in his delightful book of that title – people who solve all the world's problems in Kyoto on Tuesday (at the expense of some poor taxpayer somewhere), fine-tune their solutions on Thursday in Rio de Janeiro, take the weekend off, and then

attend a coordinating committee meeting in St Moritz, Switzerland, on the following Monday. Wouldn't their patterns of
interaction actually *increase* with distance? And you are quite
right, I suppose if we looked hard enough we could find exceptions to our distance decay rule, such as these jet setters, but
we would have to select the examples pretty carefully, and we
are looking for general regularities rather than exceptions to a
rule. And you cannot have exceptions in the first place unless
you have a rule or some regularity to begin with.

The second objection you might raise is that it takes time
and money to move over geographic space and make telephone
calls, and so rich people are more likely to move farther than
poor people. This means that if we construct a graph of the
distance decay effect for poor people it will be much steeper
than for the rich. Or, if you like, in a rich and highly mobile
society like Western Europe, North America, Japan and
Australia, the effect of distance will tend to be much less than
for a poor society like Chad, Surinam or Mozambique. Again,
you would be quite right: even within a particular society we
can see that rich people have much wider social contacts than
poor people. In Paris, for example, people in the wealthy 16th
arrondissement move much more freely to make social visits
than people living in the poorer 13th *arrondissement* (Figure
6.2), and that wedge on the map pointing to the centre of the

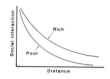

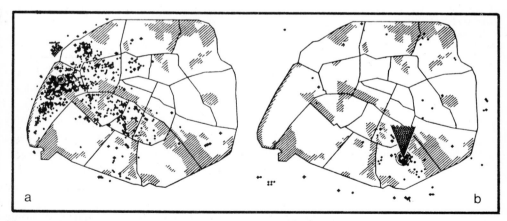

Figure 6.2: Social contacts of five wealthy families in the 16th *arrondissement*
(a) and of five poor families in the 13th *arrondissement* (b) in Paris. The
'wedge' of black dots in (b) means all the social contacts are within the
smallest inner ring over very short distances.

latter means that all those dots, all those social visits, took
place within the small circle. We just cannot show them any
other way at the same scale. So, yes, distance affects different

people in different ways, and we may very well want to take this *varying* effect of distance into account – how it changes between different groups of people, and how it changes over time.

Now it turns out that the distance decay effect, which we can interpret as the effect of geographic distance on human behaviour, is perhaps the most fundamental theoretical principle in geography. It is simple to understand, and quite in accord with most of our intuitive experiences, yet it is at the heart of most theoretical constructs in modern geography. In general, I am going to try to avoid symbols and formulas, but it is important here that we write:

$$I = f(\frac{1}{D})$$

meaning that interaction (I) is some inverse function of distance (D). In other words, as distance (D) gets bigger, so interaction (I) gets smaller. This rule holds for all sorts of things and at all sorts of geographic scales. For example, the information people have about other places tends to decline with distance away from them, and at larger geographic scales things like air and rail passenger traffic moving between cities are also subject to the same effect.

But, you may object, surely it isn't just a matter of distance, surely other things are going to affect flows of information, and how many people fly between cities – not the least of which is how big the cities are. New York and Chicago are about 1100 kilometres apart (about the same as Paris and Rome), yet they obviously interact more than New Orleans and Cincinnati (or Toulouse and Hanover), which are at the same distance from each other. The answer is, yes, of course, big places tend to generate more people, more money (bank flows), more information, almost more everything (except peace and tranquillity) than small places, so we are going to have to amend our formula and say:

$$I = f(P_1, P_2, \frac{1}{D})$$

which says interaction is a direct function of the size (P for population) of the two cities interacting, and an inverse function of distance apart. In fact, we could write:

$$I = k.\frac{P_1 . P_2}{D^\beta}$$

which says that we could predict the interaction between two cities, say New York (P_1) and Chicago (P_2), by multiplying

60

their populations and dividing by the distance between them.

Now wait a minute, you say to me, What's that k doing, and what's that Greek β (beta)? Where did they come from? Well, relax: k is just a little constant value that scales our prediction of interaction up or down, depending upon whether we are trying to estimate air passengers or telephone calls or whatever, and that β is just the power to which we might want to raise distance to adjust our formula to fit some real data or information we have. Actually, that β is really rather useful, so let us think about it for a minute. Suppose we set $\beta = 2$, so our formula becomes now:

$$I = k \frac{P_1 . P_2}{D^2} \qquad\qquad F = g \frac{M_1 . M_2}{D^2}$$

| The geographer predicting the interaction between two cities | Isaac Newton predicting the force between two masses |

Does this remind you of anything? It looks almost the same as Isaac Newton's formula for calculating the force of gravity between two bodies – the moon and the earth, or the earth and the apocryphal apple that fell on his head. And this is exactly the reason geographers call this sort of statement (a simplification and abstraction to be sure) the *gravity model*. In one form or another, sometimes rather deeply disguised, it is going to appear again and again in all sorts of different areas of modern geography, and, therefore, throughout this book. This is why I said it was probably geography's most important theoretical principle, because it pops up all over the place, sometimes when we least expect it.

So let us examine that β, the exponent of distance, a little more closely. Suppose we set $\beta = 3$, so that the inverse, or retarding effect of distance is now cubed. In our equation – our symbolic model – the populations of the two cities would now be divided by a huge number (for example, $100^3 = 1,000,000$) and this would mean that any estimate we made of the interaction (I) on the left-hand side would be cut right down. So we can say that the bigger the value of β, the smaller the interaction is likely to be. It is for this reason that geographers call β the *friction* of distance. In a sense, when β is large, the geographic space over which people or information or things move tends to be very 'sticky'. In other words, it takes a lot of effort or time or money to move over such a surface. Conversely, when β is small, the retarding effect of distance is lessened, our geographic space becomes more slippery, and people and things move much more easily. In fact, what happens when $\beta = 0$? Recall that *any* number raised to the 0

61

power is equal to 1, so whatever the distance between two places, whatever the numerical value of D we put in the equation, the populations are just going to be divided by 1. This means that there is no friction of distance when $\beta = 0$, the effect of distance has been wiped out, and people are living in a frictionless geographic space.

Is this case realistic? Well, probably not really, because we never reach a point where distance is absolutely irrelevant. But take the case of a city like New York, where a bus ride costs the same amount no matter how far you travel. If we were to measure distance in cost rather than miles, we would actually be travelling in a rather strange geographic space in which the distances (costs) between all points are the same. Or take Paris and its Metro, for which you can buy a monthly ticket (*la carte jaune*) which will let you use the Metro as much as you like for that fixed monthly fee. Again, cost is not zero, but clearly movement is partially released from the constraint of distance, although obviously the amount of *time* people have to roll around all day on the Metro is going to come into consideration.

We also know instinctively that the friction of distance has declined historically (although with increased energy costs these days it may be going up again). At my own university, the main campus at University Park draws undergraduate students from all over Pennsylvania. Around the turn of the century, the friction of distance was higher than today, so how far away a person lived was quite important in explaining whether he or she attended the university or not – the first state university in America, founded by Abraham Lincoln in 1859 under the Morrill Act to provide greater educational opportunities for all. But even Lincoln could not do much about the friction of distance in those days, although the new technology of the railway had already brought it tumbling down from the extremely high values of the canal and stagecoach days. Today, the effect of distance on university attendance is negligible within Pennsylvania, and it virtually has no effect whatsoever in explaining the pattern of students attending the main campus. Each county today, close to the main campus or far away, sends students roughly in proportion to the population living there.

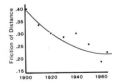

When working with the gravity model, or models of geographic interaction closely related to it, geographers may have to work with rather complex mathematics, and employ very large computers to do all the calculations. These problems are usually termed *calibration* problems, as the gravity model

62

in some form or another is fitted or calibrated to a particular set of actual data – perhaps a survey of the daily journey-to-work in a city (something the geographers on the research staff of the Paris Metro are very interested in for obvious reasons), or perhaps origin-destination surveys conducted by the Federal Aeronautic Administration for air passengers in the United States. The advantage of going to all the trouble and computing is that if we get good fits when our models are calibrated to actual data, we can turn everything around and use our calibrated model to *predict* what the interaction between places might be – at least over the short run. Let us see how this might work in quite practical and concrete terms. Let us pretend we are geographical consultants who have been asked by the Brazilian government to investigate the possibility of putting in new air links between the cities and towns of that country, or perhaps increasing or decreasing the frequency of service between towns with already existing links (Figure 6.3). Now

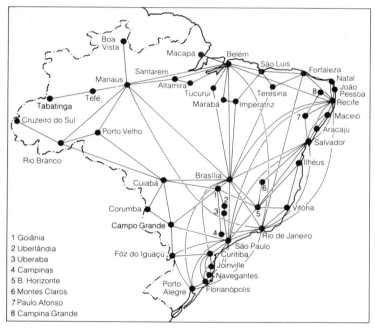

Figure 6.3: The existing internal air network of Brazil connecting 45 cities and towns whose existing flows might give the geographer-consultant important information about certain forms of interaction.

1 Goiânia
2 Uberlândia
3 Uberaba
4 Campinas
5 B. Horizonte
6 Montes Claros
7 Paulo Afonso
8 Campina Grande

no one has any *specific* theory which would allow us to predict immediately the interaction of air passengers between towns in Brazil. That is to say, we know nothing that would tell us *exactly* how Brazilians interact, as opposed to Australians, Canadians or Chinese, or the people living in any large country like Brazil. So we have got to get information specific to the

Brazilian situation, and then calibrate our gravity model to the data that are particular to that time and place. So the first thing we would ask the Brazilian government to do would be to undertake a careful origin-destination survey, probably by counting up all the actual flights people had made from the used airline tickets collected over a couple of months – perhaps one month during the summer holiday season, the other during a more 'typical' month, whatever we might mean by that.

If there are 45 towns in Brazil with scheduled services, this means we could make up a square table (a matrix) with the towns as rows and columns, and record at each intersection the number of people travelling between each pair of towns (Figure 6.3). In this way, we could collect information about the movements between nearly 2,000 possible interacting pairs, and on this basis calibrate our gravity model with great precision, finding the average friction of distance for Brazilian air travel, as well as other useful values, perhaps numbers we could use as 'weights' on the origin and destination of populations. We could also make our basic gravity model considerably more elaborate, and take into account the particular function that a city served. About 80 per cent of air traffic is business traffic of one sort or another, so two towns strongly characterized by business activities might be weighted somewhat more than agricultural service towns. Similarly, we know that cities in resort areas attract much heavier flows than usual in the form of tourists vacationing in the region. These special activities could also easily be taken into account.

Once the computer has calibrated our model, and the fit is reasonably good (as it almost invariably is, because the gravity model is very robust), we can estimate pretty closely what the potential interaction might be between any pair of towns, including those without any direct links at the moment – say Campo Grande in the southwest to Belem at the mouth of the Amazon River, or São Paulo in the south to Manaus far to the north. All we have to do is substitute in the populations, the distance and the calibrated 'weights' we have calculated on the basis of all the information we presently have, and then work out the interaction we would expect from our equation. This would help us to make recommendations about putting in a new link or not, and the size and type of aircraft to be employed (no use putting in a huge 747 between two small towns when a small short-take-off-and-landing (STOL) plane would do). We could also pinpoint new lines that might pay their way, and show that others probably did not stand a chance – no matter what the local politician and Chamber of Commerce said.

64

Of course, other questions of route *structure* and scheduling would also have to enter the advice we give, and these would influence the decisions ultimately made. Sometimes a new link can pay its way if it wiggles around a bit and picks up people at intermediate stops. One of the very early and now classic studies of route structure was undertaken in the nineteenth century, when Mexico was laying out its railway system. The question came up time and time again: how much do we deviate from the straight and narrow in order to pick up passengers and freight from intermediate towns? It is another geographic way of asking that question we had before: how do you combine things (in this case link up towns) to get the best solution when there are so many ways it can be done? It comes down to that problem of combinatorial search all over again.

We also know that adding transportation links – road, rail or air – can enhance a town's image, making it more connected, more 'central', so that it stands a better chance of attracting more people. It can also work the other way around, for new road links to out-of-the-way places may mean that people in peripheral areas will move away to the bright lights of the big central cities – as the huge rural-to-urban tides of migration attest all over the Third World. This means that there are some deep and subtle relationships between route structures and the growth and decline of towns, with each interacting in very complicated ways to produce the dynamics of a region or country. We shall take a closer look at these 'geographic systems' later, but even here we are going to find the old gravity model, that distance decay effect, embedded right at the heart of these more complex geographic systems and theories. As we have just seen, these are not esoteric theories that represent armchair speculations from the Ivory Tower. The basic theory has direct and immediate applications to many important real world problems, for there is actually nothing so practical and applied as good theory. Just how practical we shall see next when we look at human interaction over geographic space at a much smaller scale – the familiar daily journey-to-work in and out of almost every city in the world.

7 There is nothing so applied as good theory

Almost every day of the week, in almost any city of reasonable size around the world, great tides of people wash into the centre in the early morning to their places of work, and wash back again in the evening to their homes. The French have a lovely word for it, *la navette* – the shuttle that goes back and forth in a loom – but in English we call it commuting, or the daily journey-to-work. If we could station ourselves in an earth satellite high above any great city, and take time-lapse movies (say one frame every ten seconds), then we could compress a day and night into just under a minute of showing time. A year would take nearly six hours, but what a rhythmic, pulsating performance it would be. In the northern hemisphere, we would see the long wavelengths or annual rhythms as the green of summer turned to the yellow and gold of autumn, then to the white of winter, and back to the light green of spring. Every seven days we might see small explosions of people taking off for the weekend – to sand and shore in summer, and to mountains and skiing in winter – as though the city was slowly coughing and expelling some of its inhabitants.

Superimposed upon these annual, seasonal and weekly rhythms would be a still shorter one, the quick pulsation of the daily journey-to-work, the rapid panting of the city – Monday – Tuesday – Wednesday – Thursday – Friday – Pause – Pause – Monday – Tuesday – Wednesday – . . . Pant-Pant . . . and so on. Millions of people, spending millions in money, using millions of units of energy to move themselves, for we are looking at one of the most outstanding characteristics of the last hundred years – the separation of the workplace from the home. This is a very different geographic pattern from anything ever seen before, because in the old days people lived close to their work, sometimes right above it in the butchery, the bakery and the candlestick makery. Even if you worked for someone else, you certainly lived within walking distance. You had to, because unless you were rich and could afford a

66

horse there was no alternative. The friction of distance in those days was very high indeed.

Moving millions of people in two short bursts of activity every day causes enormous planning headaches involving many forms of transportation. In a city like New York, people pour into Manhattan by car, bus, train, ferry boat, hydrofoil, helicopter and underground railway, and perhaps we should not forget the few brave souls on bicycles, skateboards and roller skates! In Copenhagen and Amsterdam the proportions change, as great rivers of cyclists flow along their own well-marked paths forbidden to the bus and automobile, but the problems of congestion (and pollution) are still severe. Simply to describe such enormous systems of people moving twice each day almost boggles the imagination, for where do you start? No wonder so much urban planning is piecemeal, *ad hoc*, and uncoordinated.

The problem is that when planning takes place by fits and starts, and almost invariably in response to this year's 'sudden and unforeseen' crisis, the things we do with the best of intentions may well produce some quite unforeseen consequences in another part of the 'system'. All too often we intervene in one part of a complex system to 'relieve the pressure', only to discover that two years later things are worse than ever before. The fact of the matter is that we are not very good yet at even providing ourselves with a useful description of what is going on, let alone working out all the consequences of what we are planning to do.

Working with very large systems, in which there are lots of things moving around and interacting, is something that is rather familiar to people trying to find useful descriptions of the physical world. In particular, people in an area of physics called statistical mechanics have tried to get good descriptions of the way the trillions of atoms in a gas move around – for example, how their speeds vary as the energy available to them rises. After all, faced with trillions of atoms of chlorine bouncing around in a small cubic centimetre of a box, how do you even start to describe their movements, directions and speeds? Strange as it may seem, the analogy to millions of people 'bouncing around' a city may not be so far-fetched in terms of the basic problem of description it presents. Let us take, as a very simple example, a small 'shopping system' with ten people shopping at four stores (Figure 7.1). Suppose we keep it as uncomplicated as possible, and just record whether a person does, or does not, shop at a particular store by placing a 1 if they do and a 0 if they do not. So in our table describing

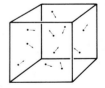

Figure 7.1: A small shopping 'system' of ten people and four stores, recording in matrix form whether they do (1) or do not (0) use the facilities.

	Vergule's Pharmacy	Michael's Meat Market	Crossroads Supermarket	Quincaill's Hardware
Scotty Alan	1	0	1	0
Godston Heronsbeach	...	...	etc.	
Robin Douglas				
Paul Nossol				
Mary Quincampoix				
Annette Seatinsea	0	1	0	0
Ann Pampilosa	...	...	etc.	
Wan Fu				
Hélène Mariposa				
Bernard Germain				

what people do, Scotty Alan shops at the Crossroad Supermarket and Vergule's Pharmacy, while Annette Seatinsea shops only at Michael's Meat Market. We record the same thing for all the people, so our table (we shall call these sorts of rectangular tables *matrices* from now on), is filled with 0s and 1s, and this particular description or configuration is obviously only one of the many possibilities, one of the many possible 'states of the system' we might observe.

Suppose I ask you *how many states* are possible, meaning how many possible ways can the 1s and 0s be arranged? You might say, well, quite a lot – let's see . . . ten people times four stores, that's $10 \times 4 = 40$ possibilities, and then each one of those can be either yes or no, 1 or 0, shopping or not shopping, so $40 \times 2 = 80$ . . . yes, 80 possible states of the system. Right?

Well, no, not exactly. You are just a wee bit off. You are quite right, we do have 40 possibilities, and each of these can be a 0 (no shopping) or a 1 (shopping), but just consider Scotty Alan's row alone for the moment. He has two choices about Vergule's Pharmacy (shop or not shop), and two for Michael's Meat Market, so that is 2×2, or $2^2 = 4$ possibilities. Then he has two choices about the Crossroad Supermarket, so that is $2 \times 2 \times 2$, or $2^3 = 8$ choices. And then two more for Quincaill's Hardware, so that is $2 \times 2 \times 2 \times 2$, or $2^4 = 16$ choices. So for

40 choices, the number of possibilities is going to be 2^{40}. And 2^{40}, give or take a few billion between friends, is just about one *trillion* possible states that our little shopping system can take. So how on earth do we even *think* about a journey-to-work involving 100,000 people going to 1,000 work locations generating $2^{1,000,000}$ possibilities, a number so huge that we cannot even say it is astronomical, because it is zillions of times larger than all the particles in the universe. It is the sort of number that mathematicians call a google, and that is all we really can do with the wretched thing – just name it, and put it back on the shelf. It is when the geographer is faced with these google possibilities that he or she decides to take up zen archery, become a poet, or spend the rest of life being a beachcomber.

Obviously, if the problem is capable of being tackled at all, it has to be approached from a different direction. And this is exactly what Alan Wilson did, because he realized that we just cannot take every person generating all those possibilities into account. *Somehow* we have got to deal with smaller numbers of things, and that means we have got to group together or aggregate people into much larger lumps, say whole residential and work areas, and deal with these instead. Suppose we had a city divided into 26 areas (Figure 7.2), and suppose that the homes of the 40,000 working people were distributed over all the areas (26 residential origins for the journey-to-work), but that only 16 of them were where people worked (16 workplace destinations), perhaps because zoning regulations forbade commercial activity entirely in 10 of the areas. We could represent our commuting 'system' as a matrix of 26 origin rows and 16 destination columns like this:

Alan Wilson
University of Leeds
1939–

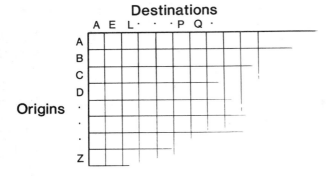

and although we sketch it here, rather than showing all the rows and columns, you can see that it still resembles the little shopping system we had before – people going to shops, and now people in residential areas going to workplaces.

So here we are with our 26 × 16 = 416 possibilities, and our 40,000 people that we have to place in their residential areas, and the same number to assign to their workplaces. Well that's a lot of good, you say to me, what on earth have we gained? You can distribute 40,000 people into 416 origin-destination boxes in your google number of ways, so we're right back where we started from! Perhaps we should take up zen archery after all? Well, it is not *quite* as bad as that. Suppose we know how many people live in each area (which is what censuses are for), and we know how many jobs are available in each work area (which is what economic surveys and industrial censuses are all about). This means that however many people we put in each little box in our matrix (say 436 people living in residential area P, and working in area K), when we have assigned all the numbers, then all the rows and columns have got to add up to the proper residential and job totals respectively. Requiring the rows and columns to add up to the residential and workplace populations means we cannot just assign our 40,000 people in any old way. For example, we cannot put 4,420 people in the 26 boxes in, say, column N, if workplace N only has 126 jobs to fill there. So by adding the information about the row and column totals we have cut down or *constrained* the possible ways people can move from residences to workplaces.

Hm . . . you say, thinking about the problem now constrained on both its rows and columns, even so, the number of ways we can assign people is still absolutely enormous. Half a google is still a google for all practical purposes, and that's exactly what we're talking about – practical purposes to help us describe and plan more effective ways of handling the journey-to-work of a real city. As far as I'm concerned that beachcomber's job still looks better than the geographer's! Of course, you are quite right: the number of possibilities *is* still googlish, but remember that when we added the information about residences and jobs the number of possibilities – the number of states the system could take – *were* reduced. Actually, they were reduced quite considerably. So perhaps if we add still more infomation, we could cut down the possible states of the system still further?

One piece of information that would be quite easy to get would be the cost of moving from each residential area to each work area, and we could show these costs between all the pairs of origins and destinations in the same sort of matrix we had before. Now intuitively I think you can see that in a very wealthy city, one where most people had a very high standard of material living, a city where the amount of money people had to spend on getting to work was only a tiny fraction of their pay cheques, people would feel free to move quite long distances. For example, some wealthy professional people live up to 150 kilometres from downtown New York, and use a train, a chauffeur-driven limousine or a helicopter to get there each day. I personally think they are crazy, but you have to admit that the friction of distance is pretty low for them. Conversely, in a very poor Third World city, where costs of daily transportation represent a big chunk of the weekly wage packet, people will try to live very close to their work if they possibly can. In a sense, the total amount of money available to the people for commuting is going to affect the distances they move. Lots of money and rich people, and we would expect a very large number of possibilities; little money and poor people, and we would expect to see many more movements constrained to short distances. We might think of money as analogous to the amount of energy available to all those atoms of chlorine gas in our cubic centimetre box. With lots of energy, many of them bounce around very quickly, hitting the walls of the box many times, and so raising the temperature. I think you can see that the statistical mechanical description is still there, just lurking under the surface.

Now I am not going to start lots of mathematical fireworks here, and pull out all sorts of rather ingenious mathematical tricks, but I would like you to accept that we can take this information and *model*, in this rather simplified and abstract way, the commuting patterns of real cities. And if we have a good journey-to-work survey (information that many cities are realizing is essential for planning their transportation systems), we can actually calibrate our model to see how well it fits the real pattern at a particular time. Again, I hate throwing mathematical formulas at you for the sake of dressing things up, but we are in the area that the mathematician-turned-geographer Alan Wilson pioneered, an area that we call entropy maximization models. The final expression would look like this:

$$T_{ij} = A_i . O_i . B_j . D_j . e^{-\beta c_{ij}}$$

and if your reaction to this is, Oh, good grief, poetry looks pretty good right now!, I cannot say I really blame you. What that subscripted and superscripted mathematical mess really means is this. We try to model a very complex commuting

71

system with as much information as we can. For example, if we can distinguish between blue-collar, white-collar and professional workers in each area – both in terms of where they live and the jobs available to them – we can build this additional information into the model, and this will constrain the possible states of the system down still further. It will still be a huge number, so we make an assumption that seems a bit strange and daring when you first meet it, but perhaps it is the only reasonable one we can make. It is simply that in the absence of any other information, we have to assume that the *actual* pattern of commuting represents the *most likely state* of the system. This means that while we know there are lots of possible states in the commuting system, even after we have placed all the constraints we can on it, some of the states appear so frequently (when we juggle our people around and assign them to workplaces to fill the jobs), that they overwhelm all the others. The less likely states are particular patterns of commuting that certainly *could* occur, but they occur so infrequently compared to the others, the most likely ones, that there is only a minute chance that our real city could be in that particular configuration of daily flows. Or so we assume – in the absence of any other concrete information we could incorporate as further constraints.

So let us take that miserable-looking equation apart, slowly and carefully, and see what it really means. We can estimate the number of trips (T_{ij}) from some arbitrary residential origin, say O_i, to some arbitrary work destination, say D_j, if we multiply them together and weight each one by a value A_i and B_j respectively, and *then* . . . aha! here it comes! and *then*, multiplying by that strange term $e^{-\beta c_{ij}}$ where c_{ij} is the cost or distance between residential area O_i and workplace D_j. Now e is just a number (2.718 . . .) that pops up all over scientific work, and it really need not detain us here. The important thing to remember is that when you have a negative exponent (that $-\beta c_{ij}$ is the power of e), it means 'one over', or $1/e^{\beta c_{ij}}$. This means we really have:

$$T_{ij} = \frac{A_i \cdot O_i \cdot B_j \cdot D_j}{e^{\beta c_{ij}}}$$

so cost or *distance* between places is really on the bottom of our equation, and it turns out that our old gravity model was there all along. I told you we would see it again. And β, of course, is nothing more than our old friend the friction of distance – the bigger β, the bigger the denominator will be, and the smaller the number of trips T_{ij} between O_i and D_j.

I think you can see that this is a much more complicated gravity model than we looked at before, and for really big and practical problems it needs a large and fast computer to calibrate it, especially to calculate all those A_i and B_j weights (one for each origin and destination area), and that friction of distance term β. But those A_i and B_j terms also give us very useful information, because it turns out that they are really measures of accessibility, telling us how 'getattable' each area is in terms of its overall accessibility to the entire system. For example, if we suppose our roughly circular city (Figure 7.2) is typical of many, with lots of jobs in the city centre, and most of the residences out in the suburbs, then the values of A_i, weighting the residential areas, will tend to rise towards the periphery of the city, meaning that people living out there will probably pay an increasing financial penalty for their relative remoteness to the job market. Similarly, a work area requiring lots of people to fill its jobs will have a large B_j value if it happens to locate in a sparsely settled part of the city, so that people have to commute long distances to it. Perhaps the factories or offices supplying the jobs in such an area will have to pay somewhat higher wages to make up for the financial loss people experience when a big chunk of the weekly pay cheque goes on commuting costs.

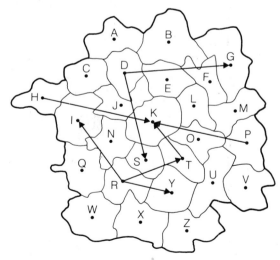

Figure 7.2: A small city divided into 26 residential areas, 16 of which are also places where people work. A few of the movements from residential areas to work places are shown, but 40,000 of these would produce an unintelligible mess.

The really surprising thing is that this essentially simple model (after all it is only the old gravity model again, even if we do give it the fancy name – entropy maximization model), does fit the empirical facts very closely indeed. For all its simplifying assumptions about most-likely states, and for all the many details we have lost in the process of greatly generalizing a real city with real people, we can achieve a very good description of the journey-to-work. And it is at this point that we can begin to play with the model. Or rather, what I really mean is use the model – I keep on forgetting as a heuristic planning device the right jargon. In fact, we do play with it, for very serious purposes, asking the sort of what-would-happen-if questions that are so difficult to answer when faced with the real world with its googles of possibilities.

For example, if we have a pretty good fit to an existing geographic pattern of commuting, we might say: suppose we put a limited access highway joining districts B-D-C-I-N-R-W on our map? This great jolt of change into the transportation system is clearly going to change the costs of moving from many of the residential areas to the places of work – and not just those areas through which the new highway happens to pass. People from A, for example, may alter their driving habits, and get on the expressway at D to take advantage of the changed conditions of *accessibility*. And this is something terribly important to remember: when we make a change in one part of a complicated system (changing just a few of those c_{ij}s in the cost matrix, for example), the effects of that change may ripple right through the entire system, altering the *relative* accessibility of some places with respect to all the others. A big change may jolt the commuting pattern completely out of whack for a while, until people learn about their changed circumstances, think about how they might take advantage of the new structure of geographic space, and settle down to a new pattern. Nevertheless, it may be possible to make some pretty good guesses about what the new pattern will look like. A geographer would say we are concerned with the *relaxation time* of the system, the time it takes to settle down after a big jolt.

Some of the changes we try to foresee may not be considered very desirable ones, and here we hit the question of values, what is considered good or bad, desirable or undesirable, and *who does the deciding* – Big Brother, or the people who are going to be affected. These sorts of questions get pretty close to home, and remember, no matter how complex this seems, we are only looking at a small part of a real city. For example, suppose oil prices rise, and costs of commuting go way up. What would we expect to see over the long run? Looking at it historically, the enormous horizontal spread of cities, and the strange things happening at the edges, have mainly taken place in this century – the century of the automobile and bus, and the enormous decline in the friction of distance for most people. Admittedly, for many cities this decline started earlier in the nineteenth century – it was when the first lines of the new Underground were built in London, with their penny fares for workers, that London started to explode outwards. But it is really the twentieth-century automobile that allows the long journeys-to-work we see today, even if it is only used to get people to the nearest commuting station from one of the surrounding dormitory towns. Notice how our language – *dormitory* towns – reflects how we think about the places we live in! It is the availability of cheap energy for transportation that has produced the sprawling modern city, and that vast space-eating thing we call Los Angeles is undoubtedly the epitome of our times.

But what happens if transportation costs go *up*, not down? What will happen to those automobile-produced sprawls then? Clearly, all sorts of things *could* happen. People might start moving back to the centre of the city if most of the jobs are there, bidding up the land prices so that only the rich can live close to the city centre. So what happens to the poor? Well, ask any old and poor person what happens when 'urban renewal' comes along. Study after study, in many cities all over the world, shows that their lives are very often destroyed. After all, planners work with abstract models and aggregated things called residential areas, not people. Or the jobs might move out of the city, leaving a great hole in the middle of the doughnut. No jobs, no tax base, at least not given the way geographic space is divided up today – eighteenth-century political divisions for twentieth-century problems.

So questions of rich and poor, fairness and equity, values and political clout enter even a highly abstract model, providing we are willing to think through some of the implications and possibilities. Even in my own little university town in the middle of rural Pennsylvania, worrisome questions of fairness and accessibility appear. Many faculty of the university and business and professional people live close to the centre, within walking distance of the campus and downtown. Many staff, service people, janitors and so on have to live farther away where houses are cheaper. Some of the poorest workers commute by automobiles over very long distances, and in the winter time the snow and ice on mountain roads means slower drives, higher costs and less time to spend with the kids. The children too pay that A_i locational penalty: up earlier in the morning for the school bus, and less chance ('the last bus has gone!') to

75

take part in extra-curricular activities such as sports and cultural events. Lots of good human things decline with distance from the centre – including opportunities for children.

So our journey-to-work is only one thread in the rich geographic fabric of a modern town or city, yet if we pull that thread carelessly it can alter the pattern of movement drastically, and also produce rather sudden change in other parts of the overall system. As we have seen, a basic key here is the change that comes about in accessibility. This business of serving and helping people by making things readily available to them in geographic space is also something of great practical importance, so let us take a deeper look at the question.

Being close to things and people

8

The idea of accessibility, of being close to things, is so much a part of our everyday intuitive experience that we seldom dust it off and examine it in any formal way, or think about it very deeply. Yet in modern geography, at both the applied and theoretical levels, questions of accessibility appear again and again. At the applied level, a great deal of geographic planning and consulting have questions of accessibility very close to the surface, while many of the theoretical aspects can be pulled together into what geographers call *location theory*. This is not a succinct theory expressed in just a few equations – geographic and human problems are much too complicated and diverse for that – but a reasonably coherent body of different ideas and methods, many of which have only become applicable in the last few years because of the computer.

Rather than plunging in at the level of complicated computer models and programs, let us start with more direct and immediate experiences. At some time or another we have all been schoolchildren, walking or taking the bus five times a week, for about 36 weeks each year, for perhaps 12 years – which comes to about 2,160 trips back and forth to school for each of us! Now it would be nice if the school were reasonably close, so we would not have to spend hours each day getting there – unlike all those poor adult commuters, who seem to spend half their lives sitting on crowded trains getting to their work and back. No wonder so many keel over at an early age – they are exhausted by all those long journeys at dawn, and all those weary trips after dark when the day's work is meant to be over. For children it is even worse: if the school is far away, it means you have to start very early on those cold winter mornings, and arrive at school already a bit sleepy. Then you have to take the bus back after school, and only then can you settle down to do all that homework before bedtime. Children seem to have so little time to be children any more, particularly those constantly chivied and organized by parents into dancing classes and piano practising and improvement activity this and improvement activity that. But that is another story. Whatever the circumstances, it does help if the school is reasonably close by.

Many parents, when they move to a new town and look for a home, take the question of school location very seriously indeed. We'd like to live farther out, they may say, but it's so far away from the schools, and Christopher and Lisa would have a long way to go. So access to schools may well be one of the most important considerations that young parents take into account. After all, the poor kids are going to have to make those 2,160 trips! But closeness to schools has even deeper implications, particularly for very young children. In the course of studying the question of school consolidation, meaning that lots of small, old-fashioned, out-of-date and inefficient(?) schools were going to be amalgamated into a few, large, new, modern, rational and better(?) ones, it was found that the distances young children travelled went up considerably. Well, obviously, you say, because if you close down four little schools, and substitute one big one, the big one is going to be farther away and less accessible to most of the children who have to use it. Yes, 'obviously', but as a result the young children may feel a deep sense of anxiety, and perhaps you yourself can recall that rather scary day when you went to kindergarten for the first time, and then your mother walked away leaving you with the teacher and all those other kids you had never seen before. It was a panicky feeling, and what we realize today is that these anxieties can affect the 'performance' (what a *word*! are we training seals or educating children?) of young children. We also realize that their intensities are often directly related to the distance the children have to travel, and whether they walk or use the bus (Figure 8.1). A young child feels secure close to home, but may become anxious as the distance increases. If the child can walk to a school not too far away, there is always the feeling that if something goes wrong he or she can get home to Mom or Dad quite easily. But if it is a long way it is not so easy, and the average ability to adjust to school goes down. Going on the bus is even worse (lower line), because there is no way children can get home on their own, so the adjustment tends to be even lower for the little ones who have to use it. So how much 'rationalization' of the school system do we really want for very young children?

The other side to this rationalization business is that thinking carefully about school districts can save a great deal of money. If the boundaries of school districts are carefully drawn, and

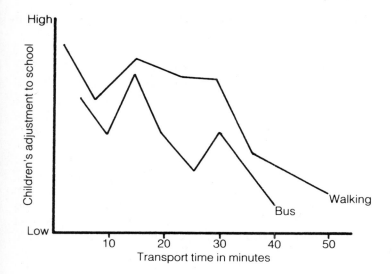

Being close to things and people

Figure 8.1: The way young children adjust to school is related to the distance they have to travel each day. The farther away, the more anxiety they tend to feel, and the more difficult it is for them to settle down.

the structure of the bus routes properly planned and considered, the school buses will only have to make relatively short trips to pick up the children. Some geographical studies have shown how many thousands of dollars can be saved each year in fuel costs. It is here that we really meet for the first time the question of how we might measure accessibility, and so adjust geographic space – the way we divide it up, locate things in it and connect it up – to give us the best arrangement, the most accessible solution to a particular problem. Obviously, at one extreme, we could have a private tutor in every child's home, and at the other extreme just a single school for everyone. In many Third World countries, for example, there may only be the resources – the money, books and teachers – to build one school to serve a rather large area. Where shall we put it? What is the *most* accessible location?

Suppose we were geographical consultants working in a rural area of a poor country, and we wanted to find the most accessible place for a new school – although you can substitute a rural infirmary, a family planning centre, a new well, an agricultural extension service, or whatever you like, because the basic problem is the same. Obviously, we would have to know where the villages are, and how many schoolchildren (or whatever particular population had to be served) there are in each one. Suppose villages A to D are located fairly close to one another, while village E, with 52 children of school age, is rather more remote. Where shall we build the school? One of the ways we could do it is by building an analogue computer – and I promise

79

you I am being serious. We could actually build a simplified physical model of the situation and use it to solve our problem. This is how we would go about it.

We take a map of the area, and glue it down on a sheet of plywood (Figure 8.2). Then, at each village, we drill a small hole and thread a smooth string through it (that slippery monofilament nylon fishing line is ideal for the purpose). We then tie all the strings together on top of the map, and underneath attach a small pan to each one into which we can place weights proportional to the number of children in each village – 45 in A, 36 in B, and so on. If we let the knot represent our school, then we can think of each village tugging the school towards it with a force equal to the number of children who are going to use it. Where the knot ends up is the best location in the sense that if all the children now walk to school, and we multiply by all the miles they walk, that total quantity will be at a minimum. Any location that deviates from this will increase the overall mileage of the children. The knot location is the most accessible according to this rather precise criterion we are using.

Figure 8.2: The analogue computer, or actual physical model, to find the best location for a school serving the children in the villages of a rural area in a Third World country.

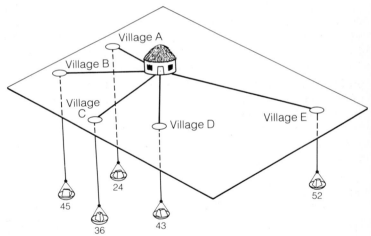

But would you really solve the problem this way?, you say to me. I mean, is this *really* how geographers go about it? Well, if we were in a poor rural area, without computers, and perhaps even without electricity, we might very well try to solve the problem in just this way. In fact we could scurry around to all the shops in the villages, borrow the weights from their scales for an hour or two, and have a big public demonstration of how we are locating their new school. In actual fact, even this simple sort of problem would be solved with a computer today, and as soon as we move to problems of more than one facility (school, infirmary, day care centre, etc.), then we would *definitely* have to have lots of computing capacity at our disposal.

There is one more question before we leave this problem of locating a single facility to serve people. It is the question of whether the location our analogue computer has selected is *fair*. Villages A to D are sitting pretty, but what about the poor kids living at E – they have a long journey to school and back every day. Could we not shift the school a little closer to them, even if the children at the other villages had to walk a bit farther? And what will happen to school attendance at E if the school is so far away? Here, of course, we meet our old friend, that distance decay effect again. It may not be a simple one in this case, described by a smoothly falling curve (recall the marginal diagram on page 59), because there may be sudden changes at certain distances – certain threshold effects – but it will be there all right, embedded as usual in our essentially geographic problem. So once again questions of fairness and justice arise in the solution of a geographic problem involving people, and if the people could discuss the location of *their* school, instead of having some outsider impose it on them, they might well decide that a location nearer E is more equitable – according to whatever intuitive values of fairness they hold.

Our simple analogue computer served us well for this problem, even though we would use a computer to solve it today, but as soon as we start thinking about locating two or more things to serve people scattered unevenly over a geographic landscape we run into really quite horrendous problems. In fact, these problems could not be solved until a few years ago, when geographers like Gunnar Törnqvist in Sweden started to tackle them with the really large machines that were making their first appearance. Intuitively, we can see why these multiple location problems blow up in our faces by thinking about a rather simple and abstract landscape like a chess board (Figure 8.3). Suppose we wanted to locate two hospitals in a region divided into 64 cells, with the number of people to be served marked in each cell (I have only put in a few of the values to keep it uncluttered, and those dots . . . mean the population values really continue over all the cells). What are the best locations – *two* places now – that make the hospitals as accessible as possible? Trial and error seems to be the best approach: we locate one hospital in the top left-hand corner, and move the second step by step to each of the other cells. Each time that second hospital makes a move, we have to

Gunnar Törnqvist
University of Lund
1933–

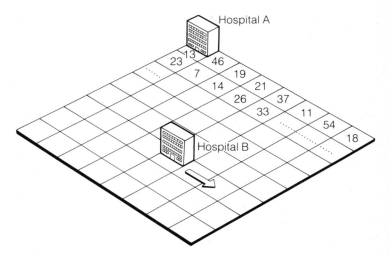

Figure 8.3: A 'chessboard' landscape with the number of people to be served by new facilities (schools, hospitals, fire stations, voting booths, etc.) in each square or cell.

calculate the cost (distance) of moving all the people in each cell to the hospital nearest them. I think you will agree that that is an awful lot of arithmetic! With the first hospital in the top left-hand cell, we have 63 other locations for the second. So that is 63 times we have to calculate the cost (distance) of moving everyone, and we would have to do this 64 times (one for each of the cells where we could put our first hospital), or $64 \times 63 = 4,032$ times. Actually, it is only half this, or 2,016, because it does not matter whether we interchange hospital 1 or 2, and we do not want to repeat a possibility we have already calculated, but it is still another of those nasty problems of combinatorial search we met before. And with 3 facilities . . . 4 . . . 5 . . . how can we *ever* find the best set of locations in real regions divided, say, into 100×100 cells?

Fortunately, the computer is very good at doing repetitive dogsbody arithmetic very fast, and I think you can see why the solutions to these sorts of problems have had to wait until the high speed computer made its debut. There was just no practical way to search for, and evaluate, all those possibilities, although today very ingenious and efficient computer programs make mincemeat out of them. In Sweden, for example, a study was made to find out the best locations for plants making cement-building blocks. Sounds pretty dull to me, you say, after all, most people don't get turned on by cement blocks. But if you think about it for a minute, locating cement block plants *is* very important. Cement blocks are used in huge quantities every year by the building industry, and they are very heavy and costly to move over long distances. Actually, in Sweden they are not so heavy, because they are often made of

82

cement foam that is almost as strong as regular cement, they are *much* lighter, and they have much better insulating properties. In countries that are falling farther and farther behind technologically they have yet to make their appearance, but you cannot keep good ideas down forever. In Sweden, it was assumed that the demand for building blocks was roughly proportional to the way the people were distributed throughout the country, and eight plants of varying size and capacity were recommended. Notice that in the more densely populated south, the plants tend to have larger capacities and they are located close together, with one of the two biggest ones right at the capital of Stockholm.

We can also turn this accessibility question around, and instead of asking what are the best locations, we can try to measure how accessible people are to existing facilities. Again, we turn to Sweden for our example, and I think it is worth pointing out, simply as a stage aside, that a lot of very practical and applied geographic work takes place in that country. Sweden has a long tradition of the Parliamentary Commission, committees that are formed to study problems when they arise at the national level. These Commissions have their own research budgets, and often ask people in the universities to help with their research skills and expertise. For example, suppose we make up a 'basket' of social goods – a doctor, a dentist, a public library, some adult education opportunities, we can specify the contents in any way we feel is appropriate – and then ask how far people in different parts of the country would have to travel to fill their basket. If we calculate these values for each grid cell on our map, we can use them as 'spot heights' to draw in the hills and valleys of accessibility – in just the same way a surveyor would draw the contour lines on a topographic map. The valleys (low values) are places where people only have to move fairly short distances to fill their social basket, while the hills (high values) represent places that are highly inaccessible to a particular mixture of social goods. You can see that in the north, with its sparse and widely scattered population, access to these things is difficult, which is precisely one of the reasons that Sweden makes great efforts to provide them at some cost of subsidization. People everywhere need decent medical and dental care, and while access can never be exactly the same everywhere, these sorts of maps highlight in a very immediate and graphic way where the major problems lie.

In all the cases we have looked at so far, we have assumed that cost is roughly proportional to distance as the 'crow flies', and careful studies have shown that when the transportation network is relatively dense, and the scale is quite large (Sweden is 1,600 kilometres north to south), this is a reasonable approximation. But there are some parts of the world where these conditions hardly hold true, and the computer programs have to be modified to solve locational problems. For example, western Scotland looks a bit like Norway, with the sea reaching into the old glaciated valleys to form a fjord coastline (Figure 8.4).

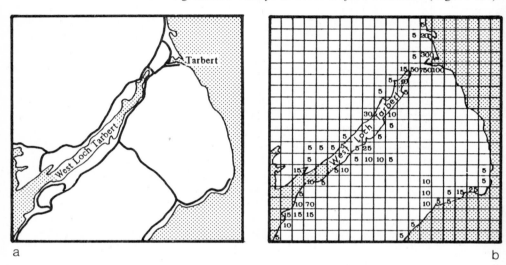

a

b

Figure 8.4: Locating four schools in a very 'awkward' geographic space in western Scotland (a), with the schoolchildren to be served located in 1 kilometre grid cells (b).

You may be able to call across to the villages on the opposite shore of the loch, or row across in a small boat, but it may be a long way by road. A schoolchild will have to take a bus, and a person who is very ill is probably much better off in a warm ambulance than in a storm-tossed boat. So how do we go about planning the provision of facilities such as schools and medical centres in a 'distorted' geographic landscape like this, where road distances may bear little relation to those covered by crows? In these cases, the computer has to be told the distances between each cell on the map over the motorable roads, and it conducts its combinatorial search for the best locations in a geographic space that has been stretched and squeezed as though it were a rubber sheet. In our particular example here, four medium-sized schools were planned to be as accessible to the children as possible (Figure 8.5), and their hinterlands, or student catchment areas, follow the roads, avoiding the uninhabited hills, and conforming to the presence of the deeply indented Loch Tarbert.

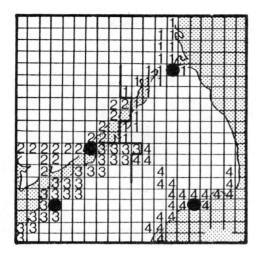

Figure 8.5: The four most-accessible locations for the schools, taking into account the indentations of the fjords and the actual road mileages. Each cell of children is assigned to the closest school, so the catchment areas and hinterland boundaries reduce the travel to a minimum.

When it comes to medical care, one of the problems we always face is that it is often needed very quickly, and today most ambulances are staffed by paramedics trained in emergency resuscitation using the latest equipment. A high level of access to such emergency services, even if it means just a few seconds of travelling time saved, can literally be a matter of life or death. So the question comes up again, where should we locate emergency ambulance services in a city to increase the chances of a person's survival? In Madison, Wisconsin, a careful study recorded the past histories of emergency cases – automobile accidents were particularly prominent – and an 'emergency surface' was drawn to show the frequency that an ambulance might be needed (Figure 8.6). With a computer doing all the combinatorial search, the geographer was able to make highly practical recommendations for the locations of emergency ambulances, and he also advised that good records should be kept and updated so that these critical services could be shifted around to better locations if circumstances changed.

No matter how hard we try, we can never create exact geographic equality, and intuitively we sense that some places are 'in the swing of things' while others are more remote – perhaps disparagingly labelled 'the boondocks'. Some people, of course, like to feel they are in the centre of things, the big towns and cities with all their variety and cosmopolitan excitement. Others (like me) would not go near them as places to live, preferring the calmer and more peaceful rural land-

Figure 8.6: The
'emergency surface' of
Madison, Wisconsin,
showing the variation
in the average number
of emergency
ambulance calls each
year.

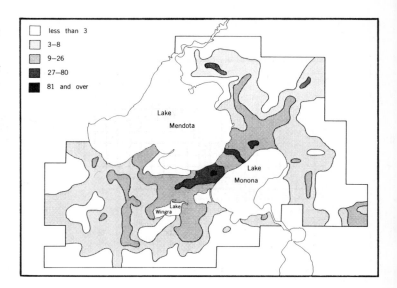

scapes. But when we use such expressions as 'being in the swing of things' and 'the boondocks' what do we really imply? Generalizations always run into the danger of being a bit too general, but on the whole I think there is some implication here of *access* to information and variety of ideas. In big cities, people tend to be exposed to lots of different people, ideas, things and information, and it is here that we often find the newest fashions and the *avant garde*, and perhaps greater tolerance towards things that are different. In contrast, small rural towns tend to be a bit 'out of it', attitudes tend to be more conservative, and people and ideas that are different are sometimes regarded with suspicion. In central Pennsylvania, for example, we can get a measure of general accessibility from our old gravity model principles – big towns lots of interaction, distant towns less interaction – and see the way the conservatism of the people varies in a quite regular fashion (Figure 8.7). These values were derived from responses to questions posed to prominent citizens (mayors, businesspeople, etc.) in the early 1970s, after all the student protests in the universities, the 'beat generation', and so on. As we might expect, those in more cosmopolitan places were far more tolerant of 'deviation' than those in small and isolated towns.

We must remember that it is not just in the developed countries that accessibility is of crucial importance. In the Third World, time can often be money too, with grave implications for the ability of a nation to feed itself – and, hopefully, others around it. Along the coastal plain of Guyana, for example,

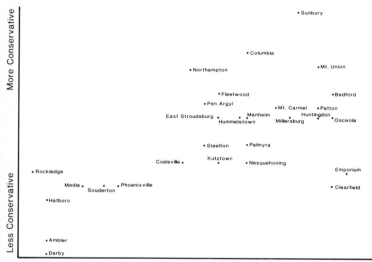

Figure 8.7: In central Pennsylvania, the more isolated a town is, the more likely its prominent people will hold to conservative and restrictive views, being less tolerant of any 'deviations' by young people from 'right and proper' ways to behave.

there are large areas that could be drained and opened up to rice farming, making Guyana self-sufficient, and allowing considerable exports to the rest of the Caribbean (Figure 8.8). A critical aspect of rice-growing in humid tropical areas is quick and efficient storage, because ideally rice should be stored within 24 hours of harvesting in termite and rodent proof facilities. In India, for example, it has been estimated that up to 40 per cent of all the grains are lost to spoilage in one form or another. So what we are talking about here is the provision of

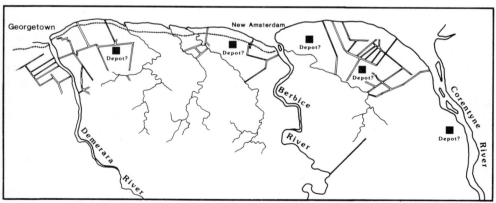

Figure 8.8: The swampy coastal plain of Guyana where plans for opening up new rice growing areas are well advanced. Rice storage depots have to be located with great care to minimize the time between harvesting and storage, and these important facilities will influence other geographic patterns in turn, for example, roads and towns.

rice storage depots in an area that is, to all intents and purposes, a new and virgin agricultural area.

A question which comes up immediately is how many modern (and therefore quite expensive) rice storage facilities do we need, what capacities should they have, and where should they be located to minimize the time and distance to get the newly harvested rice in from the fields? What we would really like are good estimates of the potential rice yields, so we could put these numbers into the cells of our chessboard landscape. Then our computer could carry out its combinatorial search for different numbers of storage depots, letting us judge the number required – a few large facilities versus more smaller ones.

What we might not realize is that we actually have a geographic tiger by the tail here. Rice storage depots along the sparsely settled coastal plain of Guyana might well serve as points of attraction for other things, in much the same way that in an earlier age of pioneer settlement a mill or a forge or an inn served as the nucleus of settlement in other parts of the world. This means that we are not just concerned with rice storage, but the small towns and villages that might grow up around them, the new roads that have to be planned to connect them up, the stages at which the new rice-growing areas are opened up . . . and so on. Everything seems to be connected to everything else, and we find ourselves right in the middle of a very big project in regional planning. At this point, we may have to concern ourselves with the goods and services (economic, administrative, educational, recreational, etc.) in our new agricultural towns, and all these responsibilities are not to be taken lightly. Investment in a town tends to be a cumulative process involving large amounts of capital that tends to have a high degree of geographic and historical inertia. Mistakes can be costly.

So in our concern for accessibility, we have touched upon questions of planning the spatial patterns and structures of undeveloped regions. Our larger concern is now the towns and villages (perhaps one day the cities?) that exist to serve people with all the requirements of modern life they can afford or think desirable. This is an area that the geographer calls central place theory, and since it forms one of the most important and fruitful areas of contemporary geography, we really ought to look at it a little more closely.

Towns as central places

9

If you think about it for a moment, the existence of villages, towns and cities – the conglomerations of things and people geographers call central places – is really rather surprising. Some of them we can explain because they traditionally served as defensive strongholds, points of religious activity, or places where one form of transportation changed to another – for example harbours and ports. But most towns just seem to happen; they just appear – rising and falling over time, sometimes explosively (Los Angeles), sometimes catastrophically (Hiroshima).

> Lo, all our pomp of yesterday
> Is one with Nineveh and Tyre!

wrote one of my favourite poets. Today, the twin central place that was called Nineveh-Nimrud is no longer, and Tyre, the major centre of the Mediterranean world about 825 BC, is a fishing village. Talk about the dynamics of central places! Even New York will look something like this one day (Figure 9.1), when the Arctic and Antarctic icecaps melt, as they undoubtedly will, whether we continue to mess about with the atmosphere or not. How nice to take a gondola down 5th Avenue.

For many geographers, trying to explain the existence of towns, and how and why they rise and fall, by advancing special and particular reasons for each case is not terribly satisfying intellectually. By that I mean that we may be able to cobble up a reason for town X springing up, and another reason for the decline of town Y, but when all the unique and special reasons have been put forward we still do not know anything of real worth about the existence of towns in general and why they change. Now a concern for general principles that in some degree illuminate an ongoing process like that of human settlements brings us into the realm of theory, and since central places are the fundamental building blocks of a human landscape, it is hardly surprising that a great deal of theoretical

Figure 9.1: New York City as the Arctic and Antarctic ice caps melt.

attention has been given to this topic by modern geographers. Moreover, the thinking in this area often tends to be rather concrete and practical, as we saw in the case of Guyana's new rice-growing areas, and the geographers who have undertaken a considerable amount of consulting in regional planning and development also tend to be the ones who have advanced our understanding of what we call *central place theory*. In actual fact, the concern for human settlements, their arrangements and their influences on their surrounding areas is an old one. As we know from our discussion of the roots of geography (Chapter 2), we have old Arab texts invoking geometric principles and patterns that look surprisingly modern. These were not developed further, however, and did not become known until quite recently. In Europe, the Englishman John Graunt wrote his *Bills of Mortality* (compiled by 'antient Matrons sworn to their office'!) in 1662, and he demonstrated an acute awareness that towns tend to form hierarchies of dominance, and that land values vary because of what he calls their 'intrinsick' properties (their fertility), and 'extrinsick' properties (due to their location relative to major market towns). In France, geographers have recently rediscovered the remarkable work of Jean Reynaud, who wrote for the *Encyclopédie Nouvelle* in 1841, anticipating many principles that are incorporated in modern theory.

It is strange how ideas come out of concealment, only to be lost again – sometimes for many generations, sometimes perhaps forever. We only have to think of Gregor Mendel, a Catholic father and the founder of modern genetics, and how his reports and papers on hybridization lay undiscovered for decades. So many original and deep scientific ideas lie in those journals of the eighteenth and nineteenth centuries, and the reports are invariably written in clear, straightforward, and often elegant prose. I remember browsing in the small library of Rutger Macklean, the father of modern agriculture in Sweden, tucked away in the beautiful old château of Svaneholm. Row upon row of bound journals from London (The Royal Society), Paris, St Petersburg . . . all carefully read and thought about during long winter evenings, all articles devoured no matter what the subject. And if they arrived two or three years late, what matter? No one was obliged to publish or perish or 'keep up' in those days.

So in areas of theoretical thought we should be wary of making claims to originality too hastily. Often, if we look carefully, we can see the faint beginnings of concern long before. But that said, there is nevertheless a clear cluster of three names in the 1930s that stands at the resurgence of central place theory as we know it today. In Germany, Walter Christaller pored over maps of the southern part of his country, and suddenly started to wonder about the rather even spacing between cities of roughly the same size (Figure 9.2). From this beginning, he developed a highly formal *geometric* theory about the way towns should space themselves apart if they were to develop on a very flat plain with roughly equal fertility everywhere so the people (mainly farmers) would also be evenly spread over the landscape. Since towns in a sense compete for the customers who will support them, big towns offering the

Walter Christaller
Independent Scholar
1893–1969

91

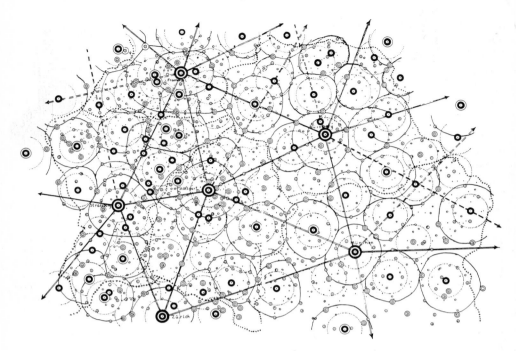

Figure 9.2: The geometric hexagonal landscape of towns in southern Germany from Walter Christaller's classic study of central places in the 1930s.

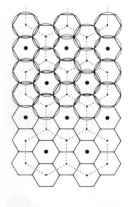

same economic functions, services and other human activities tend to push each other apart in their efforts to grab as many customers as they can from their surrounding areas. In this way, they tend to form a hexagonal lattice, since the hexagon is the geometric figure that most closely resembles a circle, but packs the space tightly without any bits left over. We can get an intuitive feel for this process of competition for geographic space by imagining what it would be like if we were locked in a room with lots of other people, and suddenly it is announced that someone (no one knows who) has bubonic plague. Immediately people look around and try to get as far away from everyone else as they can. Since everyone is trying to put as much distance between themselves and all the rest, the pattern will probably end up looking like the hexagonal lattice suggested by Christaller for his towns. We can see rough approximations to such evenly spaced lattices in Iowa, the north China Plain, and wherever rather flat and even conditions prevail in an agricultural landscape.

This essentially geometric tradition of thinking was taken up by August Lösch, whose work was translated into English just after the Second World War, prefaced by one of the most moving statements ever written by a scholar. Dedicating his book to his Swabian homeland, 'the land that I love', August Lösch seemed to know that he would never be able to explore further – 'it is not easy to stand before this rich harvest with hands tied.' He died of scarlet fever in 1945, weakened by malnutritrion, and because he was known for his anti-Nazi sympathies, he failed to gain a professorship. He gave very many provocative ideas to the next generation of geographers, and many of his insights still have great freshness and appeal.

August Lösch
Independent Scholar
1907–1945

The third name in the triumvirate is that of the geographer Edgar Kant, whose highly empirical work on towns and their zones of influence was undertaken in his native Estonia quite independently (Figure 9.3). He recognized the way large towns,

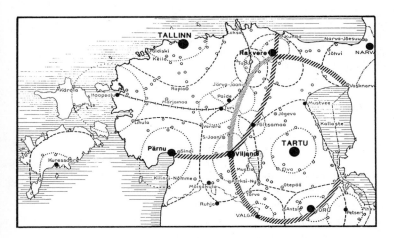

Figure 9.3: The areas of influence of Estonian towns from the early work of Edgar Kant. Notice how the central places lower down the hierarchy nestle between the largest cities (Tartu and Tallinn), yet generate their own areas of local dominance.

like Tallinn and Tartu, competed over long distances for customers, and the way smaller central places nestled between them, supplying goods and services locally for which there was a high and steady demand – essentials to life such as bread and beer. He also noticed that in between the largest towns, where their zones of influence were weakest, we tend to find 'second order' places like Rakvere, Pärnu and Viljandi. These sorts of geographic regularities were to be found again with his students in his new home in exile in southern Sweden. For Edgar Kant became *Rektor Magnificus* (bow down you mere presidents and chancellors!) of the University of Estonia, and had to go into exile when his country was taken over by the Russians.

In Sweden, his ideas and growing concern for changing central place systems were taken up by Sven Godlund, who was particularly concerned with what happened in a region where new forms of public transportation like buses were

Edgar Kant
University of Lund
1902–1978

93

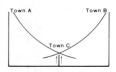

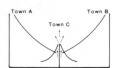

making the larger towns much more accessible to the rural people. Today we tend to forget what a new bus line meant in terms of the opportunities and changes it caused in the countryside through which it ran – increased access to a greater variety of shops, new work opportunities, museum and other educational trips for the children, and a wider area from which you could court your boyfriend or girlfriend. Over the course of half a century, the central place system of Skane (the southern part of Sweden) changed as many small places found it harder and harder to survive (a familiar story?), while a few, in rather strategic locations, managed to consolidate and draw more functions to themselves. If we imagine large towns with an urban field around them declining with the usual distance decay effect, then their influence is smallest and equal right along the boundary between them. This means smaller places have a chance to grow there, and they tend to burst through the urban fields of the larger towns like mushrooms to form little spheres of influence of their own (Figure 9.4).

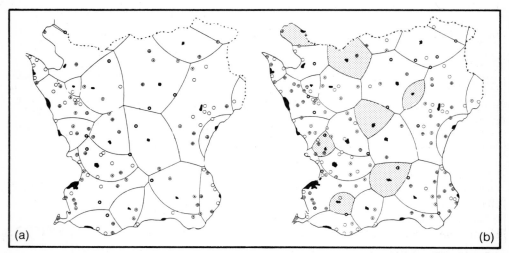

Figure 9.4: The central places of southern Sweden in 1900 (a) before the arrival of the bus. By 1930 (b) many of the small central places of 1900 lying along the boundaries have 'broken through' to form areas of dominance themselves.

In many of the original geometric approaches to central place theory, very severe, and frankly unrealistic, assumptions had to be made. Nowhere do we really find people spread perfectly evenly over the landscape, and geographic space is always bounded in some way, which causes some grave difficulties when we try to express certain aspects of these problems mathematically. Moreover, people do not just hang around a landscape devoid of towns waiting for them somehow to 'come along'. There is a highly dynamic interplay between towns growing and declining, people migrating to them from the countryside, roads being built and upgraded, and the technology of transportation. What we are really talking about is the complex mutual adjustment between patterns and flows, and the changing structures that connect them up. Sometimes geographers almost despair at handling all that interwoven complexity, and perhaps what we really need is that earth satellite and time-lapse film to animate and record the changing relationships, the flux of constant becoming. No human landscape is ever stable for long. Nevertheless, one of the really illuminating things about good theory, even if it is quite severe in its initial assumptions, is that it can always be made a bit more realistic by slowly relaxing the assumptions one by one.

For example, suppose the density of the population served by a central place system is not the same everywhere, but varies in some fairly regular way – declining, say, from the southwest towards the northeast (Figure 9.5). What would we expect the

Sven Godlund
University of Göteborg
1921–

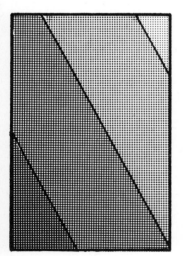
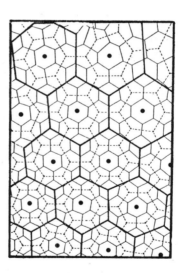

Figure 9.5: When population densities are not even everywhere, the lattice of central places adjusts to the changes, closing up in densely settled areas, and opening out in sparsely settled regions.

lattice of central places to do? In the densely settled areas, the towns can be closer together, because they can capture enough customers in the areas where they are dominant to support activities that geographers would say have quite high *threshold values* – the number of people that are required to support a particular kind of shop, a church, or even a doctor (in a style to which perhaps too many have become accustomed). But in the sparsely settled areas, the towns will tend to be farther apart in order to attract enough people to support these same activities. And give or take a little here and there, that is exactly what we tend to find in a state like Kansas (Figure 9.6), always

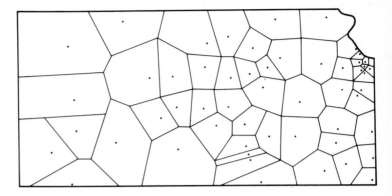

Figure 9.6: The lattice of major central places in Kansas opening out to the drier west where population densities are lower than those of the wetter east.

a favourite of beginning geography students because it is so easy to draw. Rainfall in the east is about 510mm per year, declining to about 250mm in the west. Since rain means better crops and higher profits for farmers, the farms tend to be bigger in the drier areas to generate sufficient income, so the farmhouses are farther apart, and the density of the farming population decreases. How's that for von Humboldt's environmental determinism? As a result, major agricultural service and market towns tend to be farther apart – just as our modifications to central place theory suggest. Living in the middle of the ridge and valley section of central Pennsylvania, I happen to pass through landscapes every day that are very different from those of Kansas, because all the roads, towns and people follow the long parallel valleys with their ridges in between (Figure 9.7). I have often thought that it would be rather fun

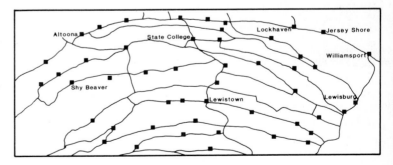

Figure 9.7: Central places in the ridge and valley section of central Pennsylvania. All the areas of influence, school districts, even marriage fields (!) tend to be elongated by the distinctive physical terrain.

two giants could pull the landscape in a northwest-southeast direction, and unfold and flatten it like the pleats in a stretched

96

accordion, so pulling the towns into a more regularly spaced pattern. However, that might be considered fudging the data to fit the theory, and besides it would turn some of my geographical colleagues apoplectic. Just between you and me, I also happen to love these hills and valleys just the way they are.

The relationships between the sizes of towns, the distances they are apart, the trade areas they serve, and the densities of their surrounding populations are not just unconfirmed speculations. If we follow the geographer Brian Berry, and examine these characteristics along a broad traverse from the densely settled area of Chicago, through the corn and dairylands of Illinois, Wisconsin, Iowa and Minnesota, on to the wheatlands of the Dakotas, and into the rangelands of Montana, we find some quite extraordinary regularities. What appears to the untrained eye as a random scatter of towns on the map becomes a highly regular set of relationships that we can succinctly summarize in graphical form (Figure 9.8). In the urban areas

Brian Berry
Carnegie-Mellon University
1934–

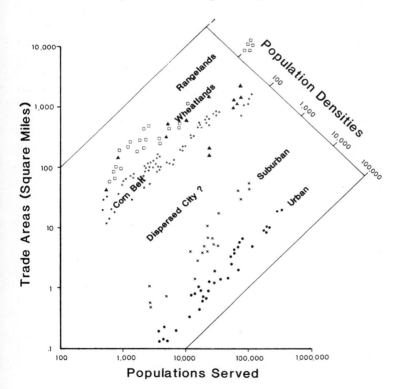

Figure 9.8: Along a traverse from densely settled Chicago to the sparse settlements of Montana, we find highly regular relationships between populations served by central places, the trade areas, and the population densities. Notice the 'gap' between the dense suburban areas and the farming landscape of the Corn Belt. Perhaps a type of 'dispersed city' belongs here?

near Chicago (large black dots), the population densities are very high, so the trade areas are relatively small, and the numbers of people served large. As we move through the suburban areas, through the corn belt to the wheatlands (open squares on the graph), the population densities decline, but the regular relationship between the trade areas and people served is maintained. Particularly remarkable is the seeming hole in the middle, where there are densities of roughly 100 to 1,000 people per square mile. What sort of central places could there be between the suburban and corn belt towns? It turns out that in southern Illinois there is a form of settlement pattern that geographers call a 'dispersed city' – as though a large city has been broken up and scattered over the landscape – and these pieces would fit perfectly into this gap.

Central place theory today has moved far from the simple geometrical, and therefore almost deterministic, models of the early pioneers. Today we can regard a human landscape as just one of the many possibilities that might have happened – once again, just one of many possible states of a system. There also seems to be a large and inherent dose of uncertainty and randomness in a central place system – no one setting up a new store has perfect information about prices and competitors, unforeseen shocks of a new superhighway through the region may occur, old industries can suddenly close down, idiosyncratic decisions to invest are made, and so on. The mathematics of central place theory tends to become more and more probabilistic (and more and more difficult!), and it has to try to tackle both discrete and finite problems (points representing individual entrepreneurs, towns, etc.), as well as those which seem continuous (urban field effects continuously declining). It is for this reason that the advanced areas of the theory today begin to look more and more like some of the problems the physicists were trying to resolve in the 1930s. Very simply, physicists faced the problem of how atoms, thought of as continuous probabilistic blurs, could radiate energy at particular wavelengths – discrete packets or concentrations of energy that could be seen as the dark lines, the sharp peaks, in their spectra. In fact, this is precisely how we estimate the different sorts of elements in stars – by looking at the spectra of their light, and finding those characteristic dark lines that say hydrogen, helium, iron, and so on. Stretching the sense of spectrum a bit, our human landscapes are also highly and sharply peaked (Figure 9.9). In Pennsylvania, for example, the wealthy urban centres form sharp peaks on the map of a generally rather poor state, and it is interesting to note that even

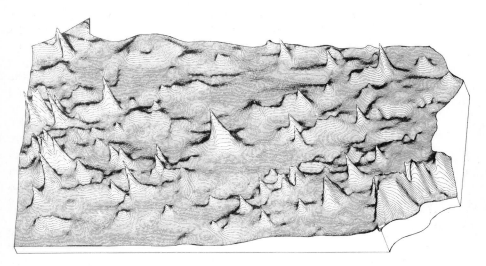

these seem to be sustained by relatively massive injections of educational money. Almost without exception, the little spikes rising up from the low plain of poverty are towns where there are colleges, universities, or other establishments of higher education. Without these continuous injections of energy, there is little doubt that the 'spectrum' would change to another configuration quite rapidly. Thus particular forces may be necessary to keep a landscape 'spectrum' in one state, and changes in the life of the people should be reflected in corresponding shifts in the geographic pattern. Leslie Curry, one of geography's major and most imaginative theoreticians today, has noted how such things as transport improvements (for example, buses in Sweden), and the disappearance of strong periodicities in work life (Monday washdays, baking days, and so on), could cause spectral shifts in the settlement pattern. It could also work the other way around: perhaps the introduction of a strong periodicity in economic activity – for example, the appearance of the sabbath day as people are converted to Christianity – could have marked effects on the pattern and growth of human settlement. This is an extraordinarily provocative idea, and one not really thinkable except from a deeply theoretical perspective that brings out of concealment possibilities not seen before.

In very old landscapes, whose configurations today have emerged over a long period of time, and whose patterns may seem a bit congealed, change may take place relatively slowly. In other parts of the world, changes in settlement patterns are going on with a quite startling rapidity, suggestive perhaps of

Figure 9.9: The socio-economic 'spectrum' of central places in Pennsylvania. Almost every spike in the spectrum represents a town with an institution of higher education.

99

A concern for theory

the jolting changes that the people themselves are experiencing. As we saw above, all these rapid changes take place as a complex process of mutual adjustment between patterns of settlements, changing connections of roads and other forms of transportation, new economic activities or the decline of old ones, changes in the rhythms of life of the people, and the introduction of new technologies – all with different lags and leads to make untangling these effects even more difficult. Geographers have long focused upon the landscape as an expression of human life in a region, but only now do our new theoretical perspectives and complementary modelling abilities enable us to move beyond an older tradition of intuitive interpretation of particular landscapes to probe some of the deeper and more general processes.

In the São Paulo region of Brazil, for example, Howard Gauthier has marshalled strong evidence for the way road investments at one time period had a marked effect upon urban growth – both population and manufacturing – at a later time. In 1940, the road network was quite sparse (Figure 9.10),

Howard Gauthier
The Ohio State University
1935–

Figure 9.10: The road network of the São Paulo region of Brazil in 1940, before hefty investments in road improvements and new road building.

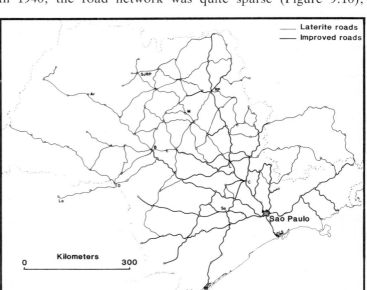

consisting entirely of laterite and gravel surfaced roads. Movement was quite costly, and the 'connectivity' of the towns, the degree of access each town had to all the others, was relatively low. But over a twenty-year period, enormous efforts were made to upgrade and extend the network (Figure 9.11), so that by 1960 the region was well served by paved roads. These

100

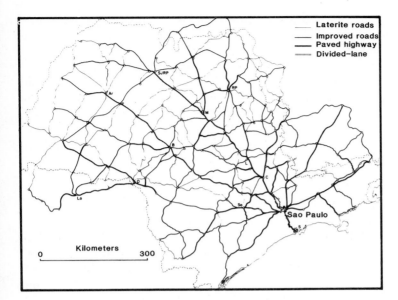

Figure 9.11: The road network of São Paulo twenty years later in 1960. These transport improvements led the rapid developments in the urban areas on the network.

greatly increased the overall accessibility of the region, and the towns that became much better connected at one time period were precisely the ones which boomed ten years later. Transportation investment clearly played a lead role, with urban activities catching up under the catalytic influence of the road improvements.

In a somewhat similar way, the French geographer Bernard Marchand demonstrated that Venezuela also experienced great jolting changes in the connectivity of its road network, because when a narrow dirt road in the interior is suddenly upgraded to a wide, hard-surfaced highway, the costs of movement are suddenly lowered. This means that in a very real sense the places being connected up move much closer together in what we might call 'cost space'. In many parts of the Third World, for example, the tarring of a laterite road can lower transportation costs to one-third of what they were before, making exports feasible for the first time as it opens up new areas to mining, forestry and farming. Transport investments produce very strange space-warping effects, and in Venezuela we can see how these result in a differential shrinking of the country (Figure 9.12). An investment that improves the accessibility of a town in the northcentral part of the country would not have much impact (short arrows), while money spent to increase the connections of a town in the east or south would rapidly shrink the space of Venezuela, bringing it close to all the other parts of the system (long arrows). Thus we can think of Venezuela

101

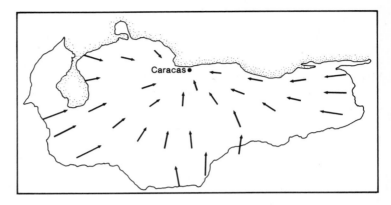

Figure 9.12: The differential shrinking of Venezuela in 'cost space' with improvements in the road transportation network. Notice how everything seems to converge upon the capital Caracas, and that the peripheral areas would be 'pulled in' very quickly by further investments compared to already quite well connected areas in the centre.

slowly converging as all the towns gradually move closer together. It is as though the map had been drawn originally on a stretched rubber sheet that is in the process of relaxing back to its original unstretched state.

With all these close and interacting effects, it is little wonder that many countries have made large investments in their geographic landscapes, putting money into those things that make up the *infrastructure* of a country – often with the help of such supranational agencies as the World Bank. If we look at the internationally supported investment programmes of many countries since the Second World War, it is remarkable what high proportions went into transportation – both main national links, and rural road developments. All these investments were made with the very best of intentions to help a new nation, and most were planned with great care for the potential impact and benefits they would bring to the people and the country. What seems to have been forgotten, with all the wisdom of hindsight that we now have, is that roads are always a two-way street – or perhaps we should say a two-edged sword. A road that links and opens up a new agricultural area, allowing new ideas, information and products to come in, also allows the people to move out much more readily. The result is that almost all over the Third World today we see great rivers of rural-to-urban migration as the bright lights of the exploding cities attract a former agricultural population like moths to a candle flame. Many delicate wings of the human condition are singed in the process. Too often the result is urban squalor on an unprecedented scale, with shantytowns and squatter areas spreading all around the major urban centres. These have grown so fast that no provision for clean water, decent shelter, adequate sewage and other basic amenities has been made, and today the World Bank has made some dramatic policy changes, shifting

102

investments in an attempt to provide at least some minimum level of planning. If ever there were an example of central place dynamics, it surely lies in these recent, and humanly distressing, shifts in the settlement pattern – that basic building block of the landscape that mirrors our human condition.

So from a concern for geographic space, in itself a rather static idea, we have started the move to a concern for time, and all the changes and dynamics produced by space and time together. I warn you, we are getting into difficult areas – difficult to think about, and difficult to handle in any reasonable way, either descriptively in words, or more formally and abstractly in mathematical terms. As a matter of fact, until very recently we could not even handle the mathematical problems, because they were what a mathematician would call 'analytically and computationally intractable'. Which means that even if we know how to write down the equations of a geographic model, we do not have the faintest hope of actually solving them in any practical and computational sense. Except . . . well, I do not want to dress up the computer in a suit of shining armour, and set him on a white horse charging to the rescue of Geographia, but it is a fact that some of these difficult problems can now be solved. The computer has really allowed us to approach problems of central place development and dynamics for the first time in a rather novel and illuminating way. Let us look at one geographic model that is causing considerable excitement for the possibilities it opens up.

10 Spatial dynamics and self-organizing geographic systems

As soon as we use the word 'dynamics' we know we are incorporating time explicitly into our thinking, looking at the way things change and develop as each new form emerges out of the last. For the geographer, this almost invariably means that we are looking at the way a map changes from one time period to another, so that a map sequence might represent a series of geographic snapshots (Figure 10.1). In peninsular Malaysia, for

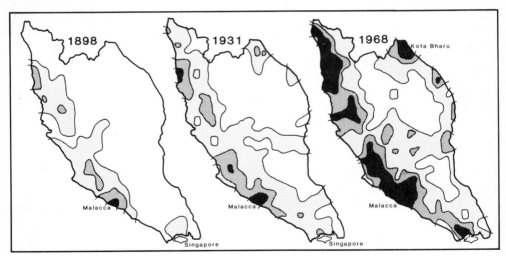

Figure 10.1: The changing road densities of Malaysia in 1898, 1931, and 1968 developing like the image on a photographic plate.

example, we can see how the patterns of road density in the later years seem to emerge out of the earlier ones. It is almost as though the most fully developed pattern (1968) was somehow latent in those of earlier times (1898 and 1931) – as though Malaya were developing like a photographic plate, so that that which is to come is somehow already there waiting to arrive out of its concealment. Geographers are fascinated by such

104

extraordinary regularities and sequences, and one of the tasks of modern geographic research – both empirical and theoretical – is to try to dig down underneath these developing patterns to see if we can find the 'rules of the game' that seem to produce them.

Geographers, of course, are not the only people who are trying to undersand how large and complicated systems like developing countries behave, and the drive to capture some of their very general properties has produced another child of this technological age – *general systems theory*. Two major figures trying to characterize very general properties of systems are Ilya Prigogine, who won the Nobel Prize for chemistry in 1977, and Peter Allen, who has worked with Prigogine in Brussels for many years. Allen started in physics, and won a Royal Society post-doctoral fellowship to work in Brussels, little suspecting that he would become a biologist, an ecologist, and finally a geographer. Yet his path to geography is really not so strange, because some of the same deep principles have been applied to modelling systems in the chemical, biological, ecological and geographical realms. The fundamental idea is that many systems are open to flows of energy, and it is these flows of energy that allow complicated systems to take on what appear to be self-organizing properties – almost as though they had a mind of their own. Sometimes the energy flows are used to maintain a system in a particular 'state', in the same way that those monetary flows for higher education enabled many of the peaks on that map of Pennsylvania (Figure 9.9) to be maintained. On other occasions, they may produce a series of developments in the system so that it reaches a point where things can suddenly flip-flop and change quickly if that point happens to be very unstable. Since this flip-flop idea is critical, let us look at it a bit more closely in terms of a system that is very important to us all – literally all around us in the air we breathe.

There is considerable concern at the moment about the increasing carbon dioxide (CO_2) content of the air, and one of the few relatively long-term sequences we have is from an observatory high above the Pacific Ocean on Hawaii (Figure 10.2). This indicates that the overall trend is definitely upwards – discounting the seasonal ups (winter) and downs (summer) caused by deciduous leaves absorbing the CO_2. Now what happens when power plants, automobiles, factories, fires . . . all the millions of CO_2 producers push more and more into the atmosphere? The answer, believe it or not, is that we still do not really know. It could increase the greenhouse effect, so the

Ilya Prigogine
The Free University of Brussels
1917–

Peter Allen
The Free University of Brussels
1944–

105

Figure 10.2:
Concentrations of
carbon dioxide
measured high in the
mountains of Hawaii.
Notice the regular
annual cycle around the
upward trend, as
deciduous trees absorb
CO_2 in the northern
summer, but fail to do
so in the winter.

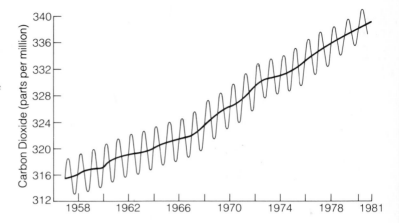

earth warms up, the icecaps melt, and most of the world's
coastal cities disappear beneath the waves (page 90). *Or* it
could increase radiation back into outer space, so the earth
gets cooler, the icecaps build up, and come rolling south again
to crush many of the world's cities beneath the glaciers. The
point is that we do not know whether the very complex
atmospheric system is in a very stable state, so that small perturbations have little or no effect, or whether it is reaching a very
unstable state – a sort of knife edge that could suddenly flip-
flop the system into another state very quickly. This is why
when three major volcanoes all go off within eighteen months
of each other, and throw up enormous amounts of debris into
the upper atmosphere, we get a bit worried about radically
altered weather patterns.

It is as though we were walking in an inebriated state along a
very broad suspended highway (Figure 10.3), and could wander
back and forth without much danger of falling off. But then
our path narrows, and just a slight tip one way or another and
we could fall from State A to State B or C if we do not make
it across the tightrope. These places where the path suddenly
becomes a knife edge are called *bifurcation points* in systems
theory, places where the system could go either way to a state
quite different from the former one – not even on the same
path or trajectory any more. The Gulf Stream, for example, is
generally regarded as pretty stable (although we do not know
its long-term relation to the heating and cooling of the earth),
even if it does wriggle around a bit from year to year. If it
reached a bifurcation point, it might change its course in a quite
radical shift, taking its warming influence away from northern
Europe, and totally changing the climate there.

106

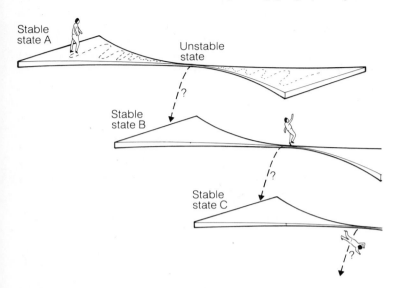

Figure 10.3: A drunk staggering back and forth along a series of suspended highways, moving from broad stable states to critical unstable ones called 'bifurcation points'.

But what has all this business about bifurcation point got to do with systems of central places?, you ask me, it seems a long way from atmospheric circulation to the dynamics of towns in a landscape! In fact it may not be such a big jump after all: a human landscape can also be considered as an open system receiving flows of energy which keep it in a particular state or configuration, and lead it along a trajectory through time to a point where it may suddenly (at least historically speaking) change. Remember Nineveh and Tyre? So what the geographer tries to do is describe a rather simple central place system by starting with a lattice of points, and then writes a rather complicated set of equations that describes how each point interacts with all the others, and how it is affected in turn by them. We are not going to write the rather horrendous set of equations here, but in essence they describe how a change in population at a particular point on the lattice is a function of ... well, quite a lot of things, all related in rather complicated ways, so we might summarize by writing:

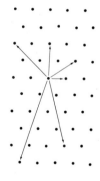

A change in population at a particular point	is a function of	the birth rate, the death rate, the rate of in-migration, the jobs available, the basic 'carrying capacity', the competition for space, crowding, unpleasantness, the ease of movement (β again!), the reaction of entrepreneurs to expand or contract, the prices of goods, the attractivity of a place, the response of the people ...

107

A concern for theory

All in all, rather a lot to handle, and many of the equations nest within each other in a sort of hierarchy of importance, and all of these have to be evaluated time and time again for each lattice point as the 'settlement history' unfolds before our very eyes. There is absolutely no way such complexity could be handled without a large and fast computer.

Yet the essence of this apparently complicated model that is going to simulate the dynamic behaviour of a geographic system is really quite simple. Economic and population growth at a point are going to take place if there are lots of jobs available, if roads make it easy to move there, if the birth rate is high, and so on (Figure 10.4). All these things have a cumulative, or

Figure 10.4: The positive and negative feedback loops that control the growth and decline of towns competing with each other in a model landscape.

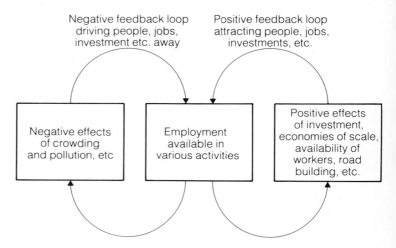

positive feedback effect, because an economically dynamic town appears attractive and draws more activities to it. However, at a certain point things get out of hand, pollution from those industries supplying the jobs makes the place less attractive, overcrowding occurs, and there is an opposite, or negative feedback effect. The history of a modern town seems to be predicated in large part on the balance between such positive and negative forces.

So how does the model work? Simply as a 'trial run', we might start with an even population at each point, perhaps a situation resembling an older agricultural landscape, and then allow each point to receive a small, 'just by chance' increase or decrease in its population. After all, the birth of a particular child, or the death of a particular person at a particular place, is not something we can really say much about or predict, so we deliberately inject a chance element into our model – what
108

a geographer would call a *stochastic* element. A computer can be programmed to produce these sorts of small random fluctuations whenever we need them. For example, if we take a really tiny system of eight points, just to keep things very simple and clear for the moment, we can get a feel for what happens over time (Figure 10.5). At first nothing much seems to happen

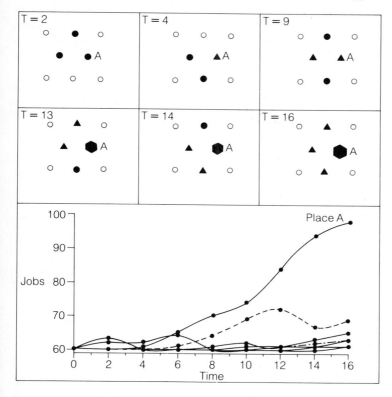

Figure 10.5: A small central place lattice of 8 points gradually developing over time, as the system becomes more and more dominated by one point that is developing into a major town.

at all: each point receives its increase or decrease by the little random injection we have incorporated into our model, and the effects of these on all the other points in the system are calculated through our set of nested equations. For a while, the little central places 'bounce around', getting a bit bigger and more attractive at one time, but being knocked down again by that stochastic shock at the next.

But then suddenly something happens, and let us follow in particular Place A. Perhaps it manages to get several additions to its population in a row *just by chance*, and in each round of calculations it begins to attract a few more people, jobs, investments and so on from the others. In a very real sense, we can say that this place has 'taken off', and it begins to impose a high degree of *structure* on the entire system. Of course, it *could* be knocked down again by a series of stochastic 'hits', but this is very unlikely now (say by generation or time 10), because it has the ability to reach out beyond itself and attract people from other places – even with a rather strong distance decay effect operating (once again, the gravity model appears). Notice how the system, taken as a whole, jogs along or bounces around for quite some time, really in a quite stable state except for small fluctuations here and there, and then quite suddenly it reaches a . . . *bifurcation point*! No more bouncing around that low level but stable state for place A: now it, and the whole system of which it is a part, flip-flops into a new trajectory leading to a new state and configuration. Now that we have the main idea, let us enlarge our example to those 50 lattice points we had before.

This larger system also 'bounces around' from the stochastic hits for the first few time periods, but by time 4 the lattice point 13 has managed to accumulate just a few more people than the others (Figure 10.6a). We shall see that this initial

Figure 10.6: After four generations of time periods (a), lattice point 13 has managed to acquire a few more people than most of the others. This initial advantage will be very important for the future development of the region, as we can see by time period 12 (b).

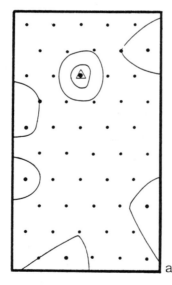
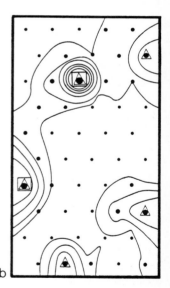

advantage is going to be very important, for once a place gets a bit of a 'leg up' it seems it can make good use of that head start, hang on to it for quite a while, and so play a prominent part in a region's development. If you think about the extraordinary historical stability of some cities, such an 'initial advantage' effect seems to be reflecting something quite in accord with our intuitive knowledge of how cities and regions develop.

110

Sure enough, by time 12 (Figure 10.6b) that point 13 (let us call it Allenville in honour of its creator) has really grown, and notice how we are representing the populations on the map – by drawing contour lines to show the changing 'population surface' of our region. A few other points are also beginning to come up, but not too close to our original one. Allenville is already quite powerful, and is 'sucking in' people and investments from nearby places before they can get going themselves. Only when the distance decay effect attenuates the power of Allenville to other point-places far away do these have a chance to grow in turn.

By time 20, our 'photographic plate' is developing nicely (Figure 10.7c), for we can see its more accentuated structure in embryonic form in the previous map. Allenville is becoming a big peak of jobs and people in the region, and may be beginning to experience urban sprawl. Also, keep your eye on that town in the southeast corner – something rather interesting is going to happen there soon. By time 34 (Figure 10.7d), those negative feedback effects of pollution and crowding are affecting Allenville, for the population at the centre of the city is now starting to decline – something that sounds very familiar – and there is the beginning of a spillover effect to the dormitory suburbs. Notice, also, down there in the southeastern part of the region, how our town has a twin rival, as these two places are competing neck and neck to become the dominant central place for that area. Finally, at time 46 (Figure 10.7e), the new upstart of the southeast has won, and the older centre falls back in the race, perhaps like an old county seat with an initial advantage that could not maintain itself against a new upstart of a university town receiving constant injections of educational funding, new students, more faculty and perhaps new 'high tech' industries congregating around it. As for Allenville, notice how that spillover effect has increased, as many surrounding towns are 'incorporated' into this urban system, serving as suburbs for the central city. Again, a familiar story?

We have here a very exciting opportunity to explore the development and changes in a central place system, and the potential has barely been touched. One of the major questions that such models raise is *how much* structure do we have to have before a geographic system will become self-organizing and begin to take off controlled by that strong positive feedback loop? What spatial structure do you need to 'get going', and, once a region is 'on the move', how much investment might you need to force it to a bifurcation point, and on to another trajectory that you consider to be more desirable? A few

Figure 10.7: After 20
time periods (c) the
region is quite well
developed with four or
five growing central
places, but by time
period 34 (d) Allenville
is spilling over into
dormitory towns, and a
rival is appearing in the
Southeast. By time
period 46 (e), this new
'upstart' has taken
over, and Allenville
continues to mature.

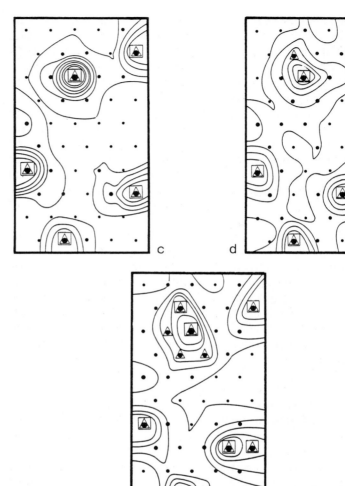

experiments show that sometimes these geographic structures
are very stable and hard to budge, and rather massive investments may be necessary to start a new and economically healthy
town. Otherwise you may end up pouring money down a rat
hole. Indeed, how many regional development and assistance
programmes have turned out to be just that?

All sorts of other possibilities arise in this particular framework of central place dynamics. For example, we can intervene
at any point we like in the 'history' of the region, and inject a
sudden change, perhaps reflecting a real jolting decision in a
real region – such as the building of a new highway, or the
start of a new industrial site. This sort of interactive process

between computer model and geographer raises some difficult questions of its own, for we somehow have to match computer time with calendar time. Computer time is really artificial (is the way we measure real time also artificial?), because we have to chop it up into 'generations' during which all those stochastic hits are assigned, our own interventions added, and the whole round of calculations made for each lattice point. But if we are trying to reproduce something close to the history of a real region, the regular intervals of calendar time may correspond to rather different intervals of computer time. It is as though one time scale were drawn on a thin rubber strip which could be stretched and squeezed differentially to fit the other. So we have to be very careful when we come to the question of applications.

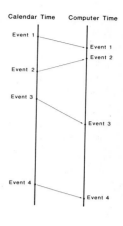

In fact, these models are so new, and have developed so quickly in theoretical and experimental terms, that very few applications exist at the moment. One, in the Bastogne region of southern Belgium, demonstrated a remarkable ability to reproduce the growth and decline of five towns in the area, but the possibilities for application at a variety of scales stand wide open. There are also some very deep ideological questions buried in these models, and it is probably fair to say that they are generally detested by geographers who look at the world through Marxist perspectives. Geographers of Marxist persuasion point out, perfectly correctly, that the model incorporates almost pure and free entrepreneurial activity, and in fact the 'self-structuring properties' would disappear – or *ought* to disappear? – under a different approach to organizing the economic activity and life of a nation. But even people in a Utopian paradise, designed by the most concerned and humane Marxist geographer, have got to overcome the friction of distance, and no matter what ideological opinions we hold, we are still in a world with much of the geographic structure already in place, produced by huge amounts of cumulative investment over decades – if not centuries. There is an enormous historical and geographical inertia built into human patterns and relationships, and sometimes these have to be taken into account. Large alterations may be very difficult to make if a system is far from a bifurcation point, unless large and wholesale shifts in people are made. Too often in history such large-scale movements have been made by force, with ultimately little effect that could be considered worth the terrible human price and consequences. Time and space are inherent properties of any human system, and we all have to deal with that distance decay effect in one way or another.

Even if we provide free buses for everyone, and try to wipe out all costs of overcoming distance, a journey still takes *time*, and time waits for no man or woman. Distance, in one form or another, still affects and structures our lives, and, as we have seen, it appears again and again right at the centre of any geographic model we can construct. That β, that friction of distance, appears at the heart of all those nested equations.

So you would think that geographers should know all about distance, what it is, how to handle it, how to measure its effects on human behaviour, and how to compare these effects from one group of people to another. Even if others forget about it sometimes – and always at their peril – surely the spatial discipline of geography ought to be right on top of this problem? Unfortunately, it is here that we face a really grave problem, with all sorts of ramifications for our geographic understanding. How do we measure the average effect of distance on people, that β value that comes into our gravity model no matter where it pops up in different guises, that distance decay effect that is so ubiquitous in our lives? It turns out that something apparently simple and straightforward on the surface can produce some first-rate geographical headaches.

Distance and the geographer's headache

11

Sometimes an area of human inquiry, like geography, economics, physics or biology, jogs along happily for years 'doing its thing', using concepts that take on the air of common possession – just ideas 'everyone knows'. Many of these are handed on from one generation to another, mainly because they have served us well, they seem quite useful, and anyway that is what we were taught in the university, so it must be true – right? Well . . . perhaps we need a bit of parody to emphasize once again that crucial difference between training and education, but it does point to the fact that too seldom do we really question those ideas that seem so obvious, the ones everyone takes for granted.

The trouble with theory, and the theoretical perspective, is that it does abstract and simplify to a very high degree, and in the course of trying to get down to the bare bones it uncovers some of these 'naively obvious' ideas and begins to question them afresh. This can be a very uncomfortable process for a discipline that thinks it knows what it is doing, but some people seem to have a special sort of 'nose' for these sorts of truly fundamental problems, and they cultivate that instinct, that intuitive feeling, and ask 'do we *really* know what we're doing?' Leslie Curry, the theoretician we met in Chapter 9, seems to have this sense honed to a pretty fine edge, for on a number of occasions he has raised questions that demonstrate that we really may not know what we are doing after all. One of these questions goes to the heart of almost every geographical model of the past few decades – the effect of distance on human behaviour, and how we go about measuring it.

Now when we talk about the effect of distance on human behaviour, what we really mean or imply is that within a group of people with roughly the same characteristics – farmers in rural Thailand, office workers commuting in Santiago, Chile, etc. – we can talk meaningfully about some sort of an 'on the average' effect of distance on people. We are talking about

Leslie Curry
University of Toronto
1922–

average effects, recognizing that there will be individual idiosyncrasies, but declaring all the same that these average effects can be estimated, even if the friction of distance will probably vary between one group of people and another. It seems to me a perfectly good and interesting question to ask: is the friction of distance less strong for Europeans compared to many people of the Third World? Or, is the commuting pattern of Athens, Greece, affected to the same degree by distance as that of Lisbon, Portugal? Or, has the effect of distance increased or decreased over a certain period of considerable technological change? Moreover, if we estimate our distance effect using data for Buenos Aires, Argentina, can we use that β value for Montevideo in Uruguay across the river? As we have seen, when gravity models, and entropy maximization models, and models of self-organizing systems (and others we shall meet later on) are calibrated, it is precisely that β term that is of great interest to human geographers. After all, human geography is about human beings moving (among other things) in geographic space. So what is all the fuss about?

Well, it turns out it is quite a fuss, and one that could take us into rather deep and difficult mathematical assumptions – and all the technical arguments that these can lead to. I want to stay away from these, but this problem is so central and important that I think we really ought to get some intuitive feel for what is at stake, even if our discussion seems to become just a bit 'esoteric'. So let us go into it in a very simple and graphical way, so we can literally 'see' what the problem is without resorting to lots of incomprehensible mathematical symbols.

Suppose we place ourselves among the gods for a moment, and make up a highly urbanized country called Linearia. Linearia is very long and narrow, so narrow that we can think of it just as a line (Figure 11.1). It lies horizontally across the

Figure 11.1: The country of Linearia, properly 'oriented' with east at the top to conform with medieval notions about what was meet and right.

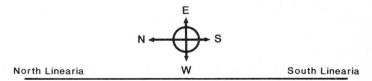

page, because a vertical Linearia would take up too much space. Only in the Chinese translation can Linearia be shown properly, and you can understand now why we know so little about the geography of Chile. It just refuses to fit on a page –

unless we cut it up, a geographic travesty that, quite rightly, upsets the Chileans. So back to Linearia – at least we have got it properly oriented.

Linearia is a highly urbanized country, and all the people live in six cities – three large ones called Alpha 1 (A1), Alpha 2 (A2), and Alpha 3, (A3), and three smaller ones called Gamma 1 (Γ1), Gamma 2 (Γ2), and Gamma 3 (Γ3). Linearia was settled in 457 BC by a group of thoroughly unimaginative three-fingered Greeks who were forced to sail from the island of Lesbos, and Alphas and Gammas were all the names they could think of when they got to their new country. The rest of the population of Lesbos stood on the quayside cheering their departure from Mytilini, said good riddance, and went back to important things like writing poetry, dancing, proving theorems and making love.

Now suppose the cities of Linearia are located at even intervals so they form a Γ A A A Γ Γ sequence from north to south (Figure 11.2), and let us also assume that there is *no*

Figure 11.2: The distribution of the wholly urbanized population of Linearia in the three Alpha and three Gamma cities.

friction of distance operating in Linearia whatsoever. In other words, the cities of Linearia interact with each other strictly in accordance to their size, and distance has no attenuating effect at all. This means our usual β would be zero ($\beta = 0$), and the Linearians interact in a completely frictionless space. Now we know, from the extensive origin-destination surveys carried out by Linearia every year, that two Alpha cities always interact at a level of 5 (5 units of interaction however we happen to be measuring it); two Gamma cities always interact at a level of 1 unit; while an Alpha and Gamma interact at a level of 3. And remember, they *always* interact at these levels, no matter where they are located, because we said distance has no effect – the usual distance decay effect has been completely wiped out.

Now suppose some poor geographer – perhaps a student writing her dissertation entitled *Towards An Understanding of the Frictional Effects of Distance on Linearian Behaviour: A Multivariate Analysis Employing Schlunk's Bifribulated Parameterization Estimates* – comes along and does not know that Linearia is a frictionless geographic space. Neither, of course, do the Linearians, because the question has never occurred to any of them during the 101 generations since the original founders of the colonization scheme. I said they were unimaginative. How would the geographer in all innocence

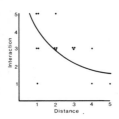

The levels of interaction between all 15 pairs of Linearian cities plotted against the distance they are apart.

proceed? As we have seen before, when we have lots of data about interactions between cities, we plot these numbers on a graph against distance, and then use various mathematical ways of estimating a best-fitting line showing us what the particular distance decay effect is. The more sharply the line drops, the greater the friction of distance, and the bigger β turns out to be.

So let us take all the 15 possible pairs of interactions between the six cities according to the annual origin-destination surveys, and plot these against the distances between all the pairs of interacting cities. With the exception of that point down in the left-hand corner (an interaction of 1 resulting from Γ2 and Γ3 close to each other in South Linearia), we find the usual downward trend, telling us that places close together interact more than places far apart. So what's new, we say, that's what the gravity model always says – interactions over long distances tend to be less than interactions over short distances.

But wait a minute, we said that in Linearia *distance had no effect!* That value of β is zero, and the people of Linearia live in a frictionless geographic space. So why does our best-fitting line have a downward slope telling us that there *is* a friction of distance when we know it does not exist? How can an effect be there when we know it is not?

This, rather precisely, is the problem we have to face here: what are we doing when we try to estimate the friction of distance from 'real' data? And, believe me, geographers do – constantly and at all sorts of scales, from the journey-to-work in a city, to air passenger movements on continental, or even world, scales. We can get a clue about what is happening if we take another possible arrangement of Linearia's cities (there are actually ten arrangements of six cities along a line we could look at), perhaps the sequence A Γ Γ Γ A A (Figure 11.3).

Figure 11.3: Another arrangement of Linearia's population in the three Alpha and three Gamma cities.

Again, there is no friction of distance, the cities interact at the same levels as before, and once again we plot the interaction level of each pair of cities against the distance between them.

But now what has happened? With the exception of that one high value (resulting from A2 and A3 interacting in South Linearia), a best-fitting line would trend *upwards*, indicating that interaction actually *increases* with distance. What a bizarre people and country – they must all be jet-setters, the international Call Girls of Arthur Koestler we met in Chapter 6 (page 58)! Now we have an opposite sort of distance effect when, once again, we know there is actually *no* effect in operation. Perhaps something is wrong? What *is* causing that distance effect when none is there? By the fact that our line on the graph changed radically when we changed the arrangement of the cities, we know it can only come from one thing: the particular *pattern* of population, or how the people are distributed across geographic space. Quite apart from distance effects due to rich people and poor people, jet-setters and stick-in-the-muds, effects that will be averaged out anyway by the methods geographers often use, the effect of map pattern is going to be convoluted in the estimate of our β.

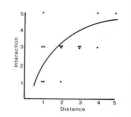

Interactions in Linearia plotted against distance when the pattern of the cities is AΓΓAA.

And to this day, we know of no way of disentangling the effect of pattern from the effect of distance.

Which is why I sometimes feel like becoming a poet or a philosopher – both, unfortunately, beyond my talents.

Well, if you are lucky, you are called to what you are. But what do people called to geography do in the face of this mess? The geographic heart of all their models of dynamic movement lies faintly pulsating at their feet. Not a terribly healthy situation, but what does it mean in *practical* terms? That is the trouble with these theoreticians, always raising problems for us practical folk who have to keep the ball rolling. Except, of course, there is nothing quite so practical as good theory . . . but we will forget that one for the moment. It turns out that for many quite practical and genuinely useful purposes, this nasty convolution of map pattern and distance decay effect does not matter too much. We can certainly use our gravity models to get rough estimates for the airlines of Brazil (Chapter 6), or calibrate our entropy models to describe the journey-to-work in a particular city (Chapter 7), and whatever values of the friction of distance we get they will certainly contribute to the accuracy of our description. The problem is that we cannot really make comparisons over space – say between the effects of distance on the commuters of Lima, Peru and Abidjan, Ivory Coast – because the *patterns* of Lima and Abidjan are almost bound to be different. Once again, we would not be comparing just the distance effects, but also the particular way residences and workplaces are arranged in those two cities – or in regions, or in nations, or at on whatever scale we are considering. This means we can never calibrate a model, say for Tokyo, and blithely apply it to Paris, quite apart from the fact that Parisians would be horrified, and absolutely refuse to be calibrated by Tokyo – or any other city for that matter!

When it comes to comparisons over time – say the frictional

119

effect of distance in Norway in 1884 and 1984 – we also have to be extremely wary. In those hundred years, Norway's pattern of population has changed quite markedly, not the least because the friction of distance has come down with the building of new roads, the introduction of new forms of transportation, and the rise in the general standard of living and material comfort. We are aware now that there is an intimate relationship between pattern and distance effects, and in Norway we know that the two different patterns of population are going to be convoluted with any estimates of the distance effects we make. We *know*, intuitively and instinctively, that the effect of distance on the geographic behaviour of Norwegians has lessened over the past century, but we have no way of really measuring it with any precision.

This problem of making comparisons between distance effects over time and space raises difficulties in many areas. For example, in northern Italy we can take groups of villages along a traverse from the high mountains, to the piedmont (literally the *foot hills*), to the plain of the Po Valley. We would expect human movement to be easiest in the valley, somewhat more difficult in the piedmont, and probably quite restricted in the steep mountainous areas. We might like to measure the effect of distance rather precisely, not only for comparative purposes, but because of the rather deep implications for studies of human genetics – genetic mixing and genetic drift (Figure 11.4). If a population, human or animal, can move

Figure 11.4: Genetic variation in Italy, as measured by the frequency of a blood type between one village and another, was greatest in the high mountains of the Alps, and declined in a quite regular way through the foothills to the plain of the Po Valley.

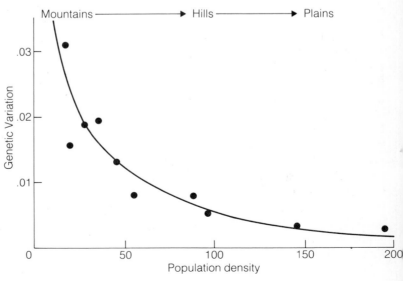

easily, the chances of genetic mixing and less inbreeding are obviously much better. In southwest England, for example, the technological innovation of the bicycle sharply lowered the friction of distance as young swains could court the apples of their eye at much greater distances, with dramatic consequences for marriage distances and less close inbreeding. But whether over space in northern Italy, or over time in southwest England, we cannot compare the pure distance effects because the geographic patterns of two people interacting also change.

Nevertheless, geographers are rather doggedly investigating this problem today, and they have turned up some quite intriguing regularities – and these are always provocative and question-inducing in a science. For example in Sweden it was found that the information people had about other places could be predicted quite well from gravity model ideas. A person tended to know a lot about places close by with large numbers of people, and not much about sparsely populated places far away. When calibrations were made all over the country, the friction of distance effect seemed to get stronger and stronger the farther away a person was from the 'centre of Sweden' – measured as the point of least travel for all the people of that country. Now, obviously, we cannot say that all the people 'in the centre' are jet-setters, feeling the effects of distance very little, and all the people at the periphery are stick-in-the-muds – or can we? No, this really will not do: the regular way the frictional effects rise with distance from the centre of Sweden

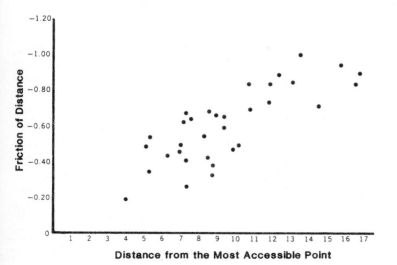

Figure 11.5: In Sweden, the calculated values of the friction of distance (β) rise quite regularly with distance from the 'centre', i.e. the most accessible point.

(Figure 11.5) *must* contain some deep aspect of geographic pattern that one day may give geographers a clue as to how they might disentangle these effects – *if* they are disentangleable at all.

It is the same in the United States. Working with air passenger movements, an area of research pioneered by Edward Taaffe, geographers have estimated these effects of distance for 100 cities involving thousands of pairs of interactions, and have found the same sort of regularities once again (Figure 11.6). In the northeast, where population densities are still very

Figure 11.6: The variation in the friction of distance (β) calculated for air passenger traffic for 100 cities in the United States.

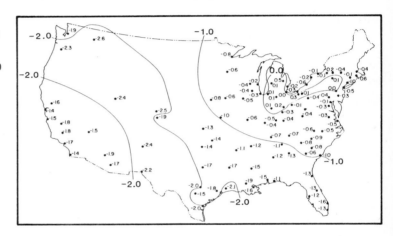

high despite a lot of out-migration over the past decade or so, the *measured* friction of distance (β) appears to be very low, and in some cases, around southern Michigan and northern Indiana, it is actually *positive*! This, of course, is totally bizarre. But as we move south and west, all the values steadily become more and more negative to the Rocky Mountain states, and then ease up again slightly when we reach California. Once again, the friction of distance is closely related to the degree of accessibility, because the most accessible point for the United States, that point of minimum aggregate travel, probably lies somewhere close to central Indiana. Somehow, in some way, the pattern of the interacting population, that *structure* of the map, is being injected into our estimations of the effect of distance on human behaviour. We do not know exactly how, and it is really a nuisance because it has stopped us dead in our tracks when it comes to measuring something quite fundamental in human geography. Of all things, we ought to be able to evaluate this important aspect of geographic space on the people who live in it.

Edward Taaffe
The Ohio State University
1921–

122

But we cannot, and if you can think of any way out, let me know. Because distance and space are not the only awkward things we have to deal with. Time is a difficult thing to handle, as we saw in the dynamics of central places (Chapter 10), and whenever human time and geographic space enter together we are obliged to start thinking very hard indeed. Let us look at some of the other ways geographers have approached these two things simultaneously.

Part III

Two perspectives: the small and the big

Human contacts in space and time

12

Whenever geographers use the words *space* and *time*, they invariably have a deeply human meaning. Glib people, who prefer scoring points rather than thinking, will immediately say that such a statement is naively obvious, trivial and banal. Time and space are human constructs, so *of course* they must have human meaning. What else could they have? But people who like scoring a point often miss it: time and space are not just fundamental, but they interact in very important ways, and form some of the most severe constraints under which we poor humans have to live our lives. Injecting an explicit and formal consideration of time into the more traditional concern for geographic space has almost become a hallmark of modern geographic studies focusing their attention on processes, developments and change. Things important for human beings happen *through* time and *over* space, and we begin to understand why geographers are fascinated by processes of spatial diffusion in which these basic dimensions both allow and constrain what we do.

As we saw with the spread of roads in Malaysia (Figure 10.1), geographers start thinking hard when they see a sequence of maps showing how the distribution of something has changed over time. Invariably their thinking (that strange sort of talking to themselves), goes something like this: Hm . . . it starts here, then spreads out . . . then it suddenly jumps to there . . . which seems to form a secondary source . . . then this spreads out until it coalesces with the original area – I wonder *why*? Now 'it' might be something as varied as a new political or religious idea, a technological innovation, or a disease moving through a population, but whatever it is we have the instinctive feeling that the movement of something over time is being channelled by the way the space is 'structured'. We will talk more about the geographer's concern for structure later, but intuitively it always involves the idea of some things being *connected* to others in certain ways. In a rather deep sense, the structure of

127

the geographic 'stage' shapes the movement of the 'players'. This is obvious when we think of a physical process, like rain-drops falling on an area highly structured by a river system. Once a raindrop flows into a tiny rivulet, connecting to a brook, flowing into a stream, that is a tributary of a river, which leads to the sea, its future movement is totally determined by the structure of the drainage network – leaving aside the possibility that it percolates into the ground, or evaporates into the air. It is not quite as simple as that in the human world, but we can often see the way human networks also shape the movement or diffusion of something. As Richard Chorley has taught us, these analogies between physical and human networks are often very helpful. In nineteenth-century Britain, for example, agrarian riots and demands for reform did not spread by a random

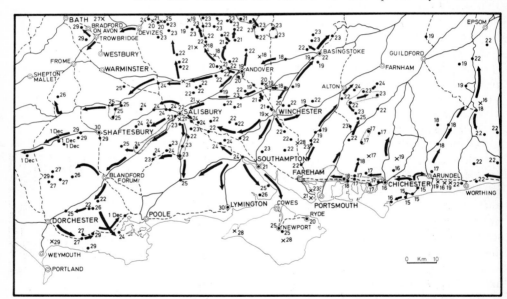

Figure 12.1: The diffusion of agrarian protest and rioting in southern England between November 15 and December 1, 1830. Notice the way the roads, particularly those used by stagecoach services, channel the protest movement.

jumping all over the map (Figure 12.1), nor did they move as a steady oozing from the original areas – the sort of wine-stain-on-a-tablecloth spreading that geographers call *contagious* diffusion. The protests were certainly channelled in part by the main road network, particularly where there were good stagecoach services carrying the news from one market town to another.

Or take another example from a very different setting: in

128

Sierra Leone, administrative reforms in the small chiefdoms were left up to the local chiefs and their councils, so there was nothing absolutely predictable about the way they diffused through the country (Figure 12.2). Nevertheless, the fact that

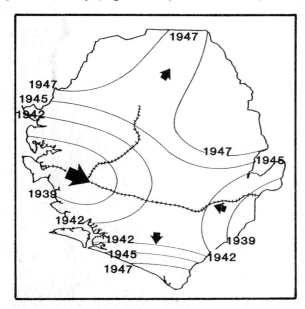

Figure 12.2: The general movement of the diffusion of administrative reforms among the traditional chiefdoms of Sierra Leone, 1939–1947. The effect of the major east-west railway line in channelling the idea is quite clear.

some areas were better connected to others by the underlying road and rail network definitely shaped the overall course of the reform movement. If the structure of the geographic stage did not absolutely determine the rate and direction of the process, like the raindrop in the river system, it did produce a strong degree of east-to-west regularity along the major railway line.

At a totally different scale, we might think of people diffusing and colonizing the far reaches of the South Pacific, sailing across this vast geographic stage on rafts carried by the currents and blown by the winds from one island to another (Figure 12.3). We could not possibly predict the exact courses of individual rafts, because we can never predict the exact strength and direction of the winds and currents, but geographers have reproduced such voyages on a computer (after all, you cannot use real rafts and real people and see whether they make it or not!), and these studies have illuminated the settlement patterns of this vast area – particularly the first settlement of New Zealand.

All of these examples make the point that in diffusion studies we may have to move away from what geographers call purely

Richard Chorley
University of Cambridge
1927–

129

Figure 12.3: The diffusion of Polynesian rafts in the South Pacific simulated by random buffeting by winds and ocean currents contained in a computer programme. Compiled from thousands of computer experiments, these indicate the drift contact probabilities of 5% or more.

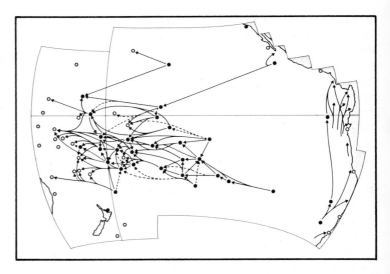

deterministic models. We simply cannot predict whether a particular person will or will not take a stagecoach on Tuesday from Winchester to Salisbury, whether they will carry the news of the recent hayrick burnings, or whether they will talk to someone who will be inspired to do the same. Nor can we tell whether a particular chief and his council will adopt the suggested reforms in March or April – or at all. Nor can we predict the way the winds will buffet a tiny Polynesian raft, or whether the people will eventually make landfall before their food and water run out. Nevertheless, we may be able to see sufficient regularity in the patterns on the maps to make some pretty good guesses. But trying to find the underlying rules of a guessing game means that we have shifted our thinking into the realm of *probabilistic* models. This is quite a shift towards complexity, and often when we try to write down the rules of the game in mathematical form the equations are so horrible that no mathematician knows how to solve them.

So what's the use of that?, you say to me. Why bother to find rules for the roulette wheel of life if you can't do anything sensible with them? Now in a quite fascinating and parallel way, this is exactly the position that two of the twentieth century's most distinguished mathematicians found themselves in around 1942. Stanislav Ulam and John von Neumann were working on the very difficult mathematical theory underpinning atomic fission during the production of the world's first atomic bomb. One of the problems they faced was how to design 'sandwiches' made out of different materials that would absorb atomic particles *diffusing* through them in order to protect the

130

people working with the deadly radioactive materials. With layers of different materials, and with rough stuff like reinforced concrete, it was not possible to follow the course of each particle and predict whether it would, or would not, get through. All you could do was follow (in a mathematical sense) thousands and thousands of particles, subject each to the random hazards of being absorbed, and count how many were likely to get through. This was really the beginning of what we now call Monte Carlo simulation, the 'Monte Carlo' indicating that sheer chance is something we must take into account, while 'simulation' means we are trying to reproduce a particular process by discovering the major rules of the game.

Now just, and quite appropriately, 'by chance', their unpublished mimeographed paper diffused into the hands of a young Swedish computer scientist, Karl Erik Froberg, who was a boyhood friend of a young geographer at the University of Lund in Sweden. As a matter of fact, most of this chapter really belongs to this geographer, for the research of Torsten Hägerstrand into the human aspects of space and time has inspired a whole generation of geographers. Starting with extraordinarily detailed information about the movements of people and the diffusion of agricultural innovations, he observed, time and again, what he termed 'neighbourhood effects' – nebula-like clusters on the map sequences indicating a strong tendency for people to adopt new ideas when they had already been accepted by others close by. Many ideas in the rural areas of Sweden seemed to spread from person to person, although it was impossible to predict if and when Farmer Cederlund would meet Farmer Paulson and talk glowingly about his new *whatchamacallit* (Swedish for bifribulated manure spreader). Nevertheless, the regularities were there in the map sequences, and it was a question of finding the rules of the game that produced them. The problem was how could you incorporate anything as slippery as 'chance' into them?

Torsten Hägerstrand
University of Lund
1916–

Suppose we make up a very simple agricultural landscape that looks like a chessboard, and into each square we put the number of people who might be *potential* adopters of an innovation (no use recording the number of males if the innovation happens to be the 'pill', or the number of adults if the innovation is a children's new 'craze'). This distribution of people is certainly going to be part of the 'structure of the space', because we would expect innovations (and diseases) to spread quickly where there are lots of people, and more slowly where there are few. We shall also have to incorporate some information about the distances over which people are likely

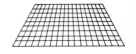

131

to communicate with each other, perhaps by asking them to keep daily diaries of their movements, or by recording the distances over which they make telephone calls. We know that, in general, the greater the distance the less likely it is that they will talk to one another, and it is this distance decay effect that we are going to record in a rather special way. We make up a little 5 × 5 chessboard, with squares the same size as the cells on our agricultural landscape, and into each cell we put the chance or probability that someone standing in the centre cell will talk to others round about (Figure 12.4a). In this particular

Figure 12.4: The distance decay effect (a) recorded as the probability that someone in the centre cell will talk to others in the surrounding cells. These probabilities can be accumulated (b) as 4-digit numbers corresponding to the chance of communication from the central cell.

.0074	.0101	.0165	.0101	.0074
.0101	.0227	.0547	.0227	.0101
.0165	.0547	.5140	.0547	.0165
.0101	.0227	.0547	.0227	.0101
.0074	.0101	.0165	.0101	.0074

(a)

0000-0073	0074-0174	0175-0339	0340-0440	0441-0514
0515-0615	0616-0842	0843-1389	1390-1616	1617-1717
1718-1882	1883-2429	2430-7569	7570-8116	8117-8281
8282-8382	8383-8609	8610-9156	9157-9383	9384-9484
9485-9558	9559-9659	9660-9824	9825-9925	9926-9999

(b)

case, there is a big chance (more than half, or 5,140 out of 10,000) of a message going to someone in the same cell, and only a very small chance (74 out of 10,000) of it going to one of the faraway corner cells. All the probabilities add up to 1.0000, because we cannot have more than certainty.

At this point we are going to make one small change, just to make it easier for our computer – which is going to do all the boring and repetitive calculations for us. Instead of putting in each cell the probability as a decimal value, we are going to put in a handful of four-digit numbers, corresponding to the chances of a message arriving in a particular cell. This sounds more complicated than it really is (Figure 12.4b). Let us take the top row, starting with the left-hand cell, where we have a probability of .0074, or 74 chances out of 10,000, that someone will communicate from the centre. In our 'four-digit' representation we shall put the numbers 0000, 0001, 0002 . . . up to 0073 into this cell – in other words 74 of these four-digit numbers corresponding to 74 chances out of 10,000. Then in the next cell, where there are 101 chances out of 10,000 (because it is a little closer to the centre), we put the 101 four-digit numbers 0074, 0075 . . . up to 0174; then the 165 four-digit numbers

132

0175 to 0339 in the next cell, and so on. We continue row by row, accumulating as we go, until we come to the last cell in the bottom right-hand corner, where we have 9926–9999.

How does this help us model a diffusion process? Suppose we assign the first two or three adopters to our chessboard landscape, the ones who first decided to try out an innovation, or perhaps those who first caught a disease. We now let our little 5 × 5 probability 'information field' float just above our chessboard, hovering over the northwest corner like a magic carpet. We then let it scan each row in turn, rather like the ball of a typewriter that returns to the left margin at the end of each row. If there is no one in the underlying cell who has already adopted the innovation, the information field continues to move along the row (Figure 12.5). But as soon as the central

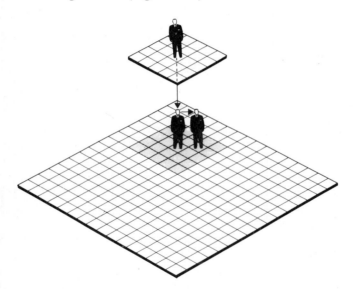

Figure 12.5: The 5 × 5 information field floating like a magic carpet over the chessboard landscape, scanning each underlying cell of population in turn, and stopping each time it comes over a cell containing an innovator or someone who has a disease.

cell comes over a square with an adopter it stops dead in its tracks. At this point, the computer generates *at random* a four-digit number. Quick! Pretend you are a computer, and give me one . . . the first that comes to mind . . . 7747, you say? OK, *where* does 7747 lie in the information field? There it is, in the interval 7570–8116 in the cell just to the east of the central cell. So on the underlying chessboard-map, we have scored a 'hit', a new person in that cell has adopted the idea – or perhaps has just caught the disease. Having generated a new adopter, our magic carpet of an information field moves on, stopping only to generate new hits, or new adopters, at each cell where some already exist.

133

Two perspectives: the small and the big

> The Moving Finger writes; and, having writ,
> Moves on: nor all your Piety nor Wit
> Shall lure it back to cancel half a line,
> Nor all thy Tears wash out a Word of it.

Persian poet-mathematicians, like Omar Khayyam, knew all about magic carpets generating the diffusion of innovations and diseases.

Some of the diffusion models are much more complicated than this, incorporating barriers into the landscape (for example long lakes or mountains that cut down the chances of a message passing), or perhaps psychological resistances ('what was good enough for Grandpa is good enough for me'). Today these complex rules are incorporated into computer programs, many of them written by Forrest Pitts, who pioneered many diffusion studies in Japan and Korea. Other models weight the probabilities in the floating information field by the numbers of people who are potential adopters in the underlying cells. This is quite sensible, because we might expect cells with lots of people to have a greater chance or higher probability of adopting an innovation, or of catching a disease. We all know that epidemics can spread like wildfire in very densely settled areas.

Forrest Pitts
University of Hawaii
1924–

Ah! But hold it right there! Do you see what this really does? It weights the probability in a cell of the information field by multiplying it by the underlying population. In other words, population is 'discounted' by distance from the central cell. Population and distance . . . sounds familiar? Of course. It is our old friend the gravity model once again, this time right at the very heart of our Monte Carlo diffusion model. I told you it would keep cropping up.

Naturally, these sorts of models must be adapted to specific conditions, and perhaps take into account other things like settlement patterns, farming rhythms and social structures. In Portugal, for example, most of the farmers do not live in farmsteads, with their farms in one contiguous piece around them, but in villages from which they go out each day to farm many different lots and fields. News about potential innovations passes over a glass of good red wine in the village square or tavern after the day's work is done – a very different process of communication compared to Sweden. Or take Mount Kilimanjaro, in Tanzania, where the Chagga farmers started a Progressive Farmers' Club, an innovation that required members to prune and spray their coffee bushes under specific rules, and carry out the initial separation of the coffee beans with great care. By conforming to these rigid requirements, they were allowed to auction their high grade coffee separately

134

from the co-operative, and so receive a higher price. Membership in the club diffused up and down the mountainside of one *mtaa* (loosely speaking a chiefdom), but the major innovator was the brother of the chief, and it was his extended family relations that provided much of the underlying structure influencing the spread of the early memberships.

In all these cases, whether it is an information field representing the average communication of a Swedish farmer, the daily ebb and flow of Portuguese farmers to the village square, or a network of family contacts in Tanzania, can you close your eyes and visualize each person, that is each one of *us* (to make it quite personal and immediate), as a point at the centre of a whole network of connections? Thinking of ourselves and other people as networks made up of both fragile and strong links is a strange way of visualizing them at first, but like many ways of 'geographic seeing' it seems more obvious and sensible the more you practise it. Perhaps you even wake up one day and cannot think in any other way. Then you know you are really hooked, and perhaps you ought to become a geographer too?

This way of thinking about personal networks has led to some very provocative and quite unexpected insights. Many of them stem from the highly imaginative work of Stanley Milgram, who most people think is a psychologist, but who we know is really a geographer at heart. He called his experiments the 'Small World Problem', because we have all had the experience of casually talking to someone we have never met before, and suddenly realizing that we both know . . . Jane Mazuski! Good heavens! Surely not *the* Jane Mazuski from Upper Lowblow? Fancy that! *What a small world it is!* What we suddenly realize is that you and I, who were complete strangers a few moments ago, are actually connected together through good old Jane (haven't seen her in years, but we still exchange Christmas cards!), so Network-me and Network-you are both part of Network-Jane (Figure 12.6). Of course, this is highly schematic and oversimplified, but perhaps you can visualize the millions of people in the country where you live as millions of little overlapping networks all connecting up into one vast tangle of communication possibilities. It is really quite a provocative, and yet deep down a very sensible, way of thinking.

Stanley Milgram
City University of New York
1933–1984

All sorts of questions pop up when we start visualizing the people of a country like this. For example, suppose you choose two people totally at random from the 220 million people in the United States, and you ask one of them to communicate

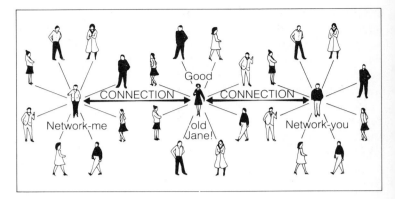

Figure 12.6: Network-Me and Network-You connected together by our mutual friend Network-Jane.

with the other, giving the name and a minimal description of who the other person is – a professor of genetics at Flotsam and Jetsam University; a woodcutter of hard maple in Sippin, Tennessee; or an abalone fisherman working out of West Clambake, California. If, just by a tiny chance, the target person is immediately known, a letter can be sent directly – a one-step connection. Otherwise the instructions say that you must send the letter to someone you know on a first name basis who *might* know the other person, or perhaps someone else who might know . . . and so on. The instructions are the same for each person in the 'chain'.

Now given two people chosen at random out of 220 million, how many people-links in the chain do you think there would have to be to connect them up? Well, who on earth could possibly say *exactly* . . . it depends, you might reply. Yes, it obviously does 'depend', but on the average, how many links would you guess? Several thousand? Several hundreds? Scores? Dozens? The actual answer, discovered by Milgram from hundreds of actual trials was . . . *five*! Most people find this hard to believe, but study after study, in various countries of the world, confirm the figure. They also confirm the existence of what we might call channels of convergence, or funnelling effects (Figure 12.7). This means that when a message comes close to a target person, it tends to be channelled through a relatively few people who are seen to be good connectors. The message of this well-established story seems to be 'It's not what you know, but who you know', distressing as this is for a democratic society striving to create equal opportunities based on merit. But from courtiers and kings in olden days, to formal computer lists of scientific networks today, access depends on structure, which is always defined by connections. In purely geographic terms, these funnelling effects mean that the steps

136

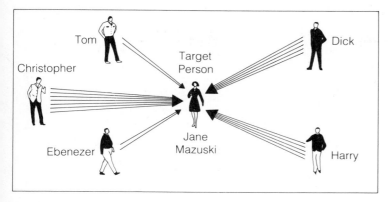

Figure 12.7: When we choose people at random to start a 'small world' chain of connection to Jane Mazuski, the messages tend to get highly channelled through those people who are perceived to be good connectors to her.

to the target get shorter and shorter on the average (Figure 12.8), and this gradual overcoming of the barrier of distance raises all sorts of exciting possibilities for measuring other kinds of barriers.

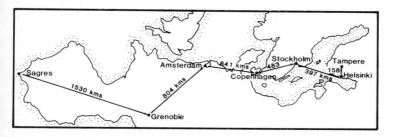

Figure 12.8: As messages to a target person get closer, the steps tend to get shorter and shorter in quite regular ways.

In the United States, for example, does it matter if a starting person is a white person, and the target is a black person . . . or *vice versa*? The answer seems to be that it makes no significant difference at all. And in Europe, what difference does it make if the starting people and targets are all in France, as opposed to France and Scotland? Or France and East Germany? How do such things as frontiers, social differences and languages chop up and splinter the human world seen as a great tangle of overlapping personal networks, a tangle that nevertheless forms a structure upon which messages are transmitted?

Messages on tangled but structuring networks are one thing, but think about yourself for a moment making direct and personal contact with other people, as well as with everyday things like post offices, grocery stores and medical clinics. After all, our normal daily experience of making contacts does not usually require us to pass messages along chains of first-name acquaintances, but consists of meeting Michael after class, having a letter weighed at a post office, buying something for dinner, and keeping that wretched doctor's appointment – why does it always have to be in the morning, so that it completely messes up my day? Answer: because the doctor has to be (he is *constrained* to be) in the surgery of a hospital 30 kilometres away between 1 and 4 in the afternoon, and so on. Do you get the idea? It is so intuitively obvious that we do not actually think about the way people's locations in time and space relative to us and each other constrain what we do. But, once again, the *structure* of geographic space, and the fact that everything takes time, allows us to do some things, and forbids us to do others. Have you ever noticed how most of us deliberately structure the little micro-spaces immediately around us, arranging the pencils, pens and paper clips of our desks, or the pots and pans of our kitchens, in such a way that the things that we need are immediately accessible to *us*. They are our own, quite personal spaces, and we structure them to suit us so that things are conveniently 'at hand'.

Unfortunately, in the larger world into which we have been thrown, we cannot arrange everything to suit ourselves. The problem is that if everything is arranged to suit *me*, then it is undoubtedly going to be most inconvenient to, and so constrain, *you*. Now the explicit examination of the constraints of time and space on people's lives has long been the subject of research of a group of Swedish geographers led, once again, by Torsten Hägerstrand of Lund. Some of the hallmarks of this type of research are the two- and three-dimensional diagrams that weld the dimensions of geographic space and time together (Figure 12.9), although these are used simply as aids to thinking, rather than total reproductions of the extreme complexity that actually makes up a modern society. Each person traces out a path in space and time, and many human projects require that these paths are coupled together. In a sense, these 'coupling tubes' are like the waist of an hour glass through which all the possibilities of the future are constrained down to one actual present.

Suppose, for example, we follow someone who is snoozing peacefully in a nice warm bed at home at 7 in the morning (Figure 12.10). The alarm clock goes off, he turns it off . . . just five more minutes. But . . . oh good heavens, it's 7.30, and I have to be at the office at 9 . . . no time for breakfast, I'll snatch a cup of coffee at the cafeteria nearby. So our hero makes it to the office on time, works there until 11, and then has to leave for an appointment on the other side of town at 11.30 . . . should have left earlier, dammit . . . better take a taxi. The discussions go on until 12.15 . . . we can walk to the
138

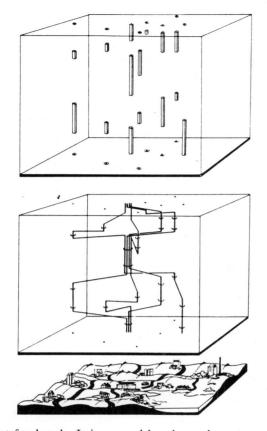

Figure 12.9: People's movements over a geographic landscape and through time. Each person traces out a 'life line', some of which pass through 'tubes' representing obligations to be at a certain place for a certain time. Many of the individual life lines are coupled together in these tubes as people meet on a face-to-face basis to discuss things.

restaurant for lunch. It is a good lunch, perhaps too good . . . 2 o'clock already? I've got to fly . . . taxi again to the office . . . work there until 5 . . . walk home, must get some exercise . . . whew, what a day . . . TV and bed.

What our hero has actually done, of course, is trace out 12 hours of his complete life-line in time and space, where vertical lines in the diagram mean he is staying in one place, while slanted lines show that he is moving to new locations that actually structure the space around him – the things and people he needs. His 'reach' in the space depends on what time he has available, and on what form of transportation he can use (Figure 12.11). Various forms – walking, bicycling, car, etc. – give him greater and greater freedom for other contacts. In fact, the opportunities available to us determine the space-time *prisms* that we live in, and it is not so far-fetched to read that as *prisons*. In a very concrete and immediate sense that we have all experienced, we are prisoners to a greater or lesser degree of the space-time prism-prisons we live in. Even these

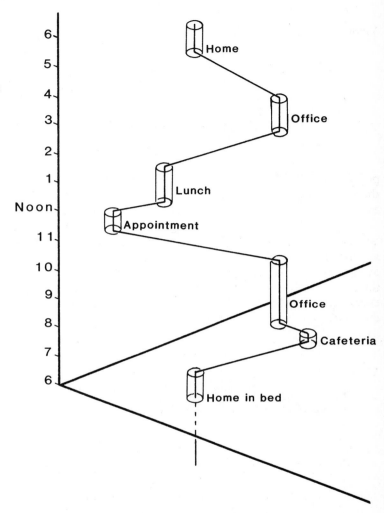

Figure 12.10: The single life line of our hero as he wakes up in the morning, somehow makes it through a gruelling day, and finally makes it home in the evening.

wedges in space-time are often cut down still further, because we often travel to meet and talk to someone else, or transact some sort of business, like buying stamps at the post office. That means if I am at X, and you are at Y (Figure 12.12), and we agree to meet at Z, our requirements for contact are so coupled together that we act as constraints on each other. As for buying those stamps, I drive to the post office, only to find that the post office does not exist any more. Doesn't exist? Don't be silly, there it is before your eyes! Yes, a physical building with Bureau de Poste written on it is there, but a *real* post office, a splendid British, American or Canadian post office with open, capacious and welcoming doors, with real

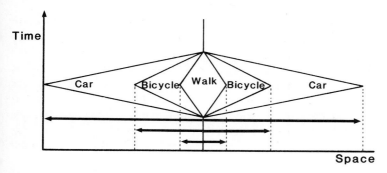

Figure 12.11: Our 'reach' in space and time depends upon the form of transportation we have available to us, and these various possibilities determine the space-time prisms we live in.

people behind real counters selling real stamps . . . that isn't there any more. Why? Because it is 12.15 and I am in France, and the French, like a sacred religious rite unfathomable to heathen North Americans, close down their public (?) facilities between 12 and 2 for lunch. The French, of course, think this is quite rational, and it is those barbarians from the other side of the Atlantic River who are mad. The barbarians in turn think the French must be insane to 'destructure' their space every day by making their Bureaux de Poste disappear from the space-time world in which they live.

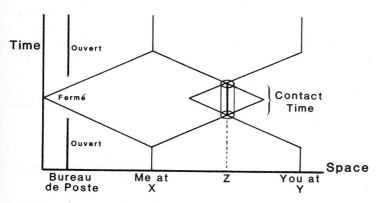

Figure 12.12: The space-time prism-prisons of You and Me determine the contact time we actually have to talk or transact some business.

Can you visualize the strange, pulsating ballet we all dance, on a space-time stage whose props appear and disappear? No wonder it is so difficult to fill our 'shopping baskets', when all the things we need, and all the people we want to meet, are scattered all over the place, opening and closing, available and unavailable. Looking at the world in this strange way, where people and things trace out life-lines constantly coupling and uncoupling, perhaps we begin to understand a little better some of the things we take for granted.

For example, only the other day, in Grenoble, France, I bought fresh fish, studded snow tyres, champagne, potting soil, electrical fuses, insulated snow boots, fresh flowers, a book on recent social history, milk, vegetables, meat and other groceries, and photographs that had been developed since my last visit – all at the same *hypermarché*, with its 100 check-out counters ringing up the francs. It took me about three-quarters of an hour. Not long ago I would have had to visit the fishmonger, garage, wine merchant, garden store, electrician, shoe store, florist, bookseller, dairy, greengrocer, butcher, grocer and photographer. How long would it have taken me? Who knows – all day?

Providing, and here is the hitch, providing I am a rich person, who lives in a rich society, with a *car*. A car not only to get to the shopping centre, but to . . . well, have you ever tried carrying fresh fish, studded tyres, champagne, potting soil . . . ? But do you also see how a technological innovation like the automobile can change not simply the movement of the dancers, but the structure of the stage? And in very poor countries, where it takes the women and children four hours to fetch water, at the same time that firewood has to be found, at the same time that the fields have to be weeded, at the same time that the goats have to be tended, at the same *time* . . . but seldom in the same *place*. No wonder people are needed to get the chores done and the family going. We are all dancers in time: only the stages are different. Every opportunity fulfilled, every contact and conversation made, is an opportunity lost, and a word that was never said. In the space-time dance we sense the constant loss and achievement of our own humanity. Not a bad thing for *human* geographers to do every once in a while.

Macro-geography: centres and peripheries

13

In the examples we have looked at so far, we have ranged from problems in very small areas (the space of your desk, the geographic effects of bubonic plague in a closed room), to those on a regional scale (systems of central places, the diffusion of innovations, the planning of new agricultural areas). The geographic scale has been from small to medium, or, if we want to sound a bit more scholarly, from the *micro* to the *meso* scale. The Greek roots always sound much more respectable. But there is also a way of looking at the world in great geographic 'chunks' that we can legitimately call the *macro* scale, where we really back away and try to see broad spatial trends and regularities – a sort of bird's eye view that we get today from earth satellites. Now I certainly do not want to draw sharp distinctions between these various scales, and I am not going to get excited if one person's upper-meso is another's lower-macro, but for me that high-level or macro perspective has a very particularly geographic 'feel' to it. And by that I mean it is a way of looking and thinking at a scale that only geographers in the human sciences provide.

Down at the micro level, I often feel I am working alongside people like psychologists, sociologists, anthropologists, archaeologists and even economists. And geographers do work, of course, with other scientists, often in very fruitful and enriching partnerships. I personally like carrying out research with people in other fields, and I have learnt a lot from them across such varied subjects as philosophy, art education, Meso-American archaeology, mathematics, health administration, nutrition, physical education and geology. This is a common, but important experience of many geographers. You can often get a good idea of the scale at which these collaborative studies take place by looking at the graphics in the publications. They are often line diagrams of things like kinship relations, graphs

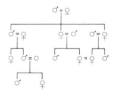

143

of REMs (rapid eye movements) plotted against arousals (psychologists are nice, but strange people), or maps of smallish areas – archaeological sites, social visits in a town, etc., all using maps at a fairly large scale.

And just as a reminder, geographers measure the scales of their maps with a ratio or fraction which tells you that one centimetre on the map represents (let us say simply as an example), 100 centimetres, or one metre, on the ground – or 1/100. Now if one centimetre on the map is equal to 100 centimetres on the ground, this is the sort of scale we would use for very detailed maps of small, or micro, areas – like a plot of ground where we are going to build a house, an archaeological site, or perhaps a map of a Balinese vegetable garden. This scale or fraction of 1/100 would be numerically much larger than the scale of a map on which one centimetre represented 10,000,000 centimetres on the ground, or 1/10,000,000 – the sort of scale a home atlas would use to show a big, or macro, area like all of Europe. The English, and others like the Australians who have had the misfortune to be influenced by them, continue to use such unwieldy fractions as 1/63,360.

Now 1/10,000,000 is a much smaller fraction than 1/100, so small-*scale* maps show big or macro areas, while large-*scale* maps show small or micro areas. Small is big, big is small, micro is large-scale, macro is small-scale . . . you can see what a confusing world geographers live in.

At the macro level the maps in the publications tell a different story, and you can be almost certain that you are looking at the work of a geographer, or the work of people who were strongly influenced at some point in their careers by geographers, like the historian Fernand Braudel. We are not exclusive or selfish about our macro-geographic viewpoint, and we encourage everyone to see the world in this way. We even welcome penitent geologists to the club – after they have seen the truth of continental drift, which geographers supported all along, despite the geologists' ungentlemanly antagonism, not to say outright derision at times. Perhaps continental drift took the geologists so long to 'see' precisely because they were always hammering their rocks down there at the micro level, so they never looked up to see the big picture right in front of them. Today, we have actually measured the rates at which the continental plates are drifting (1.5 centimetres per year for North America drifting away from Europe, and 7 centimetres per year for Australia moving towards Hawaii). It is naively obvious to any geographer that South America fits into Africa, and threw up the Andes like snow in a snowplough when it drifted away; that North America did the same with the Rockies; that the two sides of the Red Sea scrunch together perfectly; that India shot northwards and piled up the Himalayas in front of it, and so on (Figure 13.1). Of course, I am

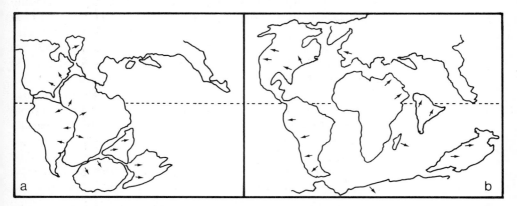

a | b

only pulling the legs of the geologists, because scepticism is always valid in a science, and it took enormous amounts of evidence, patiently gathered and pieced together, to fill in the details. Interestingly, the change in opinion really started during the International Geophysical Year of 1967–8, when international collaboration on a large scale involved maps (and their macro perspective) at small scales. For all the tongue-in-cheek and teasing, there is some truth in that business of backing away and moving up to the macro level to see the larger picture more clearly.

Figure 13.1: The theory of continental drift requires a macro-geographic perspective to see how Africa and South America fitted together (a) 180 and (b) 60 million years ago, and how such mountain chains as the Rockies, the Andes and the Himalayas were thrust up by the movements of the continents.

In human geography, this sort of backing away always means that many local details are blurred and drop out of consideration. It also means that lots of information is treated in large clumps or aggregates. The whole question of what happens when we aggregate information from the details of the micro level to the broad scales of the macro level is something we will have to think about again (Chapter 25). All we need to realize here is that as we take bigger and bigger geographic chunks as our building blocks, we will get smoother and smoother representations; 'smoother' in the literal sense that all the rough spikes and edges are rubbed off. Let us look at a concrete example.

Suppose, in the United States, we lump all the people in a county together, and simply represent them by a number at a point. We now ask, what is the potential interaction of these people with all the others in the United States? – where all the

rest are also represented by numbers at the centre points of their counties, all 3,105 of them. We know enough now to realize that we are back in 'gravity model country' again, because, in general, groups of people will interact with each other in direct proportion to their size (P), and in inverse proportion to the distance they are apart (d). So for our potential interaction measure in a single county, we simply add up all the potential movements predicted by the gravity model with all the other counties. That is 3,105 multiplications, 3,105 divisions, and 3,105 additions, but all these are done in a blink of an eyelid these days. However, we have to repeat these calculations for each of the 3,105 points or counties, so that comes to 9,315 × 3,105 operations, or 28,923,075 – once again, just a loosening up exercise before breakfast today for our computer.

So now we have potential interaction values of our 3,105 counties, and we can use these as 'spot heights' to draw our potential interaction map (Figure 13.2), giving us quite a

Figure 13.2: The map of potential interaction based on the 3,105 county values for the United States.

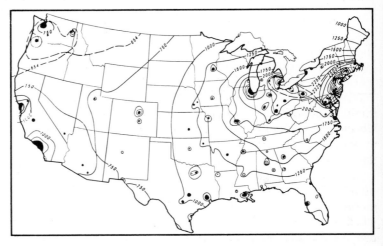

detailed, yet still macro, overview of . . . well, of what, *exactly*? We have created a 'gravity model' map, but what does it mean? It really gives us a view of the potential interactive energy of a population, and the remarkable thing is that so many other things seem to be related to this rather fundamental surface. If we were to plot movements by automobile, railway and air – of people, of mail, commodities, etc. – we would see an almost exact match, with intense flows in areas of high population potential, and weak flows in others. Notice how these broad plateaus and individual spikes delimit the major areas of commerce and manufacturing. In fact, in one sense the map is

146

too detailed, and all those 'spikes', while interesting, are really local 'noise' obscuring our appreciation of the main national 'signal'.

We can suppress the local noise by aggregating our populations into larger lumps, perhaps into the 48 contiguous states rather than the 3,105 counties, only this time we are also going to weight the population values by the average income of each state, to produce a much smoother income potential surface (Figure 13.3). On this we can see some of the gross differences

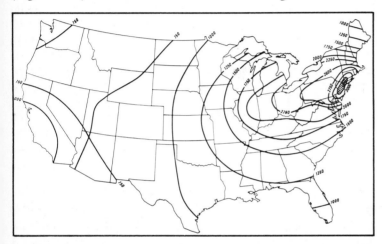

Figure 13.3: A smoothed representation of the potential interaction surface of the United States based on the 48 values of the contiguous states.

in the United States, in particular the change between the Northeast and Midwest manufacturing belt and the low values to the west and south. Now one of the remarkable things about this map, originally constructed by William Warntz, is that it is not completely stable. Income levels in regions rise and fall, people move, and over time even this very generalized view of income potential slowly changes. In particular, look at that gradient to the south – does it remind you of something? Where else do you see 'pressure' surfaces at these macro scales? On weather maps, of course, and so Warntz made an extraordinarily provocative analogy between the steep gradient of income potential, and the warm and cold 'fronts' that such gradients imply on a surface showing air pressure.

William Warntz
University of Western Ontario
1922–1988

Now weather fronts move, so Warntz calculated the income 'pressures' and gradients for successive censuses, and found that the income front was rolling south at a steady rate. Extrapolating from this, he made one of the boldest and longest-ranged forecasts that I know of in the entire human sciences – where long-range prediction is notoriously difficult, and where even demographers (population experts) hedge their predic-

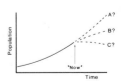

147

tions with such wildly different outcomes that they can never be said to be wrong. He predicted (and remember this was long before people were talking about sunbelts and frostbelts in America), that the income front would continue to roll southwards, and that by the year 2000 it would have wiped out most of the differences in income potential. Today, the income front seems to be just about on schedule, with all the political, as well as economic, implications it carries with it.

This sort of macro perspective has recently been given a new spurt by the migration predictions of Peter Allen, whose model of central place dynamics we met in Chapter 10. I find it very interesting that people who eventually find their way into geography from physics (or who work with physicists as Warntz did with his colleague John Stewart), seem to be able to 'get up' to the macro perspective so much more easily than many others. Following similar gravity model principles, Allen has made a number of predictions for migration in the United States that are much closer to the actual figures than all the detailed calculations of the economists and demographers. Perhaps there is something about the macro perspective after all?

The maps of William Warntz and John Stewart were very important for the way they gave human geographers a renewed sense of the broad viewpoint, and Warntz did not confine his thinking simply to human affairs. He also examined some of the ways people and their technology interact with their physical environment. Airplanes flying the Atlantic from New York to London (Figure 13.4), do not take a straight line, even allowing for different east-west routes and different altitudes to prevent collisions. Airline pilots, and those who help in planning each

John Stewart
Princeton University
1894–1972

Figure 13.4: The tracks of trans-Atlantic aircraft take into very careful account the pressure surfaces and winds, and subject to safety regulations they attempt to make maximum use of them.

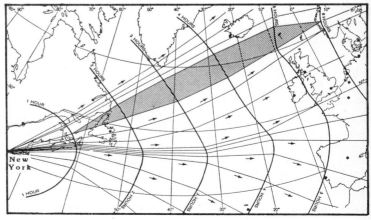

flight, are well aware that the pressure surface over which they fly can assist them. Allowing for a twist due to the earth's rotation, winds flow down a gradient at right angles to the slope (like water running downhill), and if you can wiggle your route around to catch a nice tailwind you will go faster *and* save fuel – sometimes a matter of make or break importance for airlines these days. But this close relationship between pressure surfaces and movements cuts both ways. This means you may know the surface, and so work out the best movements, *or* you may know what the movements are, and then 'work backwards' to derive the pressure surface that caused them.

Some of the most provocative and exciting work in this area of 'reverse thinking' has come from Waldo Tobler, and we shall meet him again when we look at the explosive developments in cartography (Chapter 17). As we saw in our example of planning new airlines for Brazil (Chapter 6), flows between places (perhaps people migrating between states), can be recorded in square tables, or flow matrices. Starting with these, and using ingenious computer programs to do the millions of calculations and all the graphics, Tobler has constructed maps of 'winds of influence' showing the changing forces at work. For example, suppose we ask what are the winds blowing people around when they have just received their doctoral degree and are looking for their first job (Figure 13.5)? Imagine

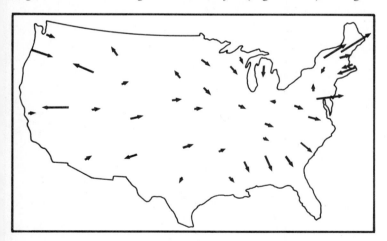

Figure 13.5: The 'winds of influence' blowing new PhDs to their first jobs. Take a person with a new doctoral degree at any point in the United States, and the direction and force of the wind blowing there will tell you the most likely way they will move.

it this way: go to any point in the United States, take a new PhD, and throw him or her up into the air – like the sort of blanket-tossing game Eskimos play. They may object, but take no notice. After all, this is all in the cause of Science. There she goes . . . up . . . up . . . which way will the winds of

149

influence blow her? If we are tossing PhDs in the northeast of the United States, there is little question where she will go – straight south to the sunbelt, and all the new opportunities it represents, pushed by a strong wind (the strength is proportional to the length of the arrow) to the hot swinging cities of the New South, and away from the cold, depressed cities of the Old North. Naturally, we will not be able to predict every individual movement, like the migration of Mary Jane Tomberlero, with her new PhD in entomology, going out to the National Caterpillar Research Station in Wahway, Idaho, because our aim is to see broad average tendencies, not the Mary Janes who are aggregated with all the other PhDs. But by and large . . . well, you get the idea.

Such maps of winds of influence can also disclose pockets of instability in the life of a nation, places where things have not 'settled down' yet in the course of a long historical trend. In Switzerland, for example, Waldo Tobler and his colleague Guido Dorigo have worked with huge flow matrices of movements of people between the 3,037 *Gemeinden*, and have found that the migration pressure surface is quite smooth, producing

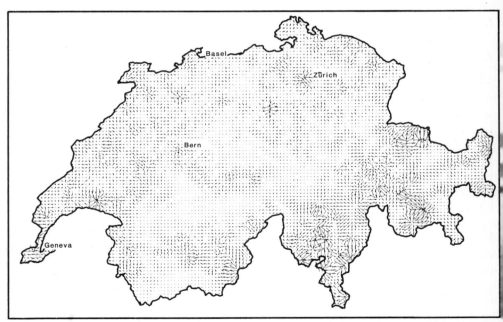

Figure 13.6: The winds of migration in Switzerland, showing small 'zephyrs' of influence over most of the country, except where dislocating changes are still going on in the southeast, where the 'migrational turbulence' is quite evident.

150

rather small and stable winds over most of the country (Figure 13.6). But in the southeast corner of the country, the pressure surface roughens, producing a great deal of migrational turbulence. What is going on here? What is the meaning of those swirls and eddies in the normally smooth flow? It seems that the older, more highly industrialized parts of Switzerland have settled down to some sort of rough equilibrium, as people move back and forth between the major towns in broad and quite stable streams. But in the southeast, people are still migrating down from the Alpine villages to the towns, changing age-old patterns of agricultural work for those of an industrial nature. The swirls and eddies disclose a macro overview of the turbulence and dislocation in individual people's lives, and perhaps we are seeing the last of a 'modernization front' that has swept across Switzerland over the past century or more, producing similar pockets of turbulence as it passed, but leaving behind the gentler migrational winds we see over most of the country today.

This rather strange, but imaginative way of summarizing huge quantities of flow information uncovers some major high and low pressure zones in the human geography of large areas. In a rather carefully qualified sense, the high-pressure areas are the centres, while the low-pressure areas are the peripheries, and I say 'rather carefully qualified' because I do not want to imply that nothing goes on at the peripheries. As we know from our weather analogy, winds of considerable force can blow in low-pressure areas too. But the concept of centre and periphery can be very helpful when we think at the macro-geographic level, particularly for people who live in parts of the world all chopped up by national boundaries, and whose thinking often seems to be rather confined to what goes on inside the separate boxes. Once again, it is that business of backing away to see yourself in the larger picture.

Simply as one example, I have the distinct impression that it is sometimes easier for Americans to 'see' Europe as a whole than it is for the Europeans themselves. Despite Common Markets and NATOs, and a mobile young generation, an awful lot of 'partitional thinking' about Us and Them goes on in Europe, and these mental images work their way up to the top to be translated into real economic and political constraints. Take a simple, everyday, but quite concrete, example: when you buy fresh California vegetables in a Pennsylvania supermarket, you wonder why such a big bureaucratic and form-filling fuss is made getting the same sorts of crisp vegetables from Portugal to Sweden. The distance is roughly the same,

151

and so is the number of states or countries the vegetables have to pass through (Figure 13.7). Those United States of Europe really ought to get their act together. If we (Americans) were doing it, we would quickly organize the vegetable market in *Europe* (and notice, *not* in Portugal, Spain, France, Belgium, Holland, West Germany, Denmark and Sweden) in no time at all, just as we do back home in the *States* (and notice, *not* in California, Nevada, Colorado, Kansas, Missouri, Illinois, Indiana, Ohio and Pennsylvania).

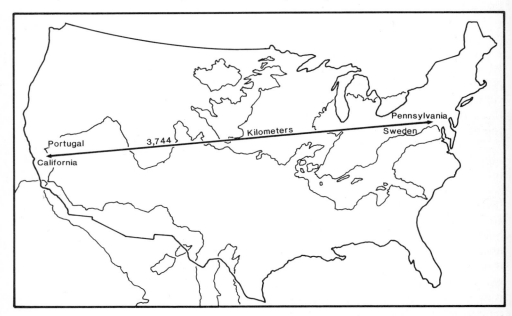

Figure 13.7: Moving fresh vegetables from California to Pennsylvania in the United States is about the same as moving them from Portugal to Sweden in Europe, with roughly the same number of state or national boundaries to be crossed in each case.

Quite naturally, there are few things that infuriate Europeans more than having Americans point up their inefficiencies, but since I live on a small rock exactly halfway between the two continents I can see both points of view – and receive the worst criticism from both directions. So let us 'think European' for a moment, wipe away all those boundaries (where are they on those satellite photos?), and see what the economic potential surface of Europe is like (Figure 13.8). Where is the 'core' of Europe? Right there in the centre of intense activity in the Paris-Düsseldorf-Frankfurt triangle. The high intensity area 'slops over' the Channel to include southeastern England, but it is really all downhill to the periphery – Portugal, Sicily,

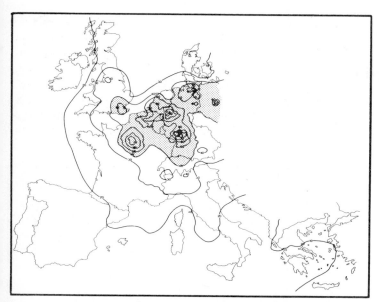

Figure 13.8: The economic potential surface of western Europe, showing the Paris-Düsseldorf-Frankfurt core, with gradients down to the periphery in all directions.

Greece, Eire and so on. Does the surface *really* match activity levels? Suppose we look ahead (page 209) at a remarkable satellite photograph of Europe, taken at night (Figure 18.1). Obviously, there are some 'outliers' – the Po Valley of northern Italy, the Glasgow-Edinburgh axis in Scotland, the Danish-Swedish twin of Copenhagen-Malmö (we shall look at them again, Chapter 23), and so on. But if ever there were a core, it is surely that bright splodge of activity in the northwest corner of the continent, and if ever there were a periphery it is those large dark areas of Spain and Portugal, southern Italy and Eire. Centre and periphery, yes; but also rich versus poor? Innovators and laggards? Well-connected cosmopolitanism, and inaccessible backwaters? *Power*?

Many things seem to vary with the same intensity as this 'pressure surface' in Europe, and for all the wars and destruction that have rent the continent over the past 120 years, it seems to have an extraordinary stability. With slight variations, we can see it appear again and again, even if we go back to almost the beginning of the last century. It is true that the railways started in Britain (Figure 13.9), but they quickly jumped the Channel to the highest point on our stable economic potential surface, and from there oozed across the map of Europe in a remarkably steady way down the gradients to the peripheries. Here and there small pockets of innovation appeared ahead of the advancing wave, but generally the fit is remarkably good. Should we be surprised? Perhaps not really.

153

Figure 13.9: The
diffusion of the railway
in Europe from the old
and stable core to the
periphery.

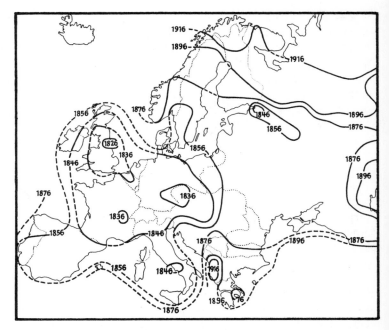

After all, trains move people and their goods, so areas of
high potential interaction attract investments quickly, and when
their potential is fulfilled, the investments tend to shift to areas
of less intensity, and so on literally down the line.

One thing we must remember, however, is that our potential
interaction values always contain two elements: population
itself (unweighted, or weighted in some fashion), and that
discounting effect of distance. Obviously population can change
as people are born or die or migrate, but that distance effect
may also change over time. As we know, costs of overcoming
distance have been greatly lowered over the past 200 years,
which means geographic space has been slowly but consistently
restructured. This means, in turn, that the influences of the old
centres may reach out and affect more strongly the places that
were truly marginal in an earlier day.

In France, for example, the TGV (*très grande vitesse*), the
fastest train in the world, has compressed the travel time
between Paris and towns like Grenoble, Chambéry and
Annecy, close to the French Alps, to just over three hours.
This means that the effect of distance in this particular direction
is greatly lowered, so the potential interaction with the great
mass of Paris goes up. You can see the effects everywhere: in
the advertisements of daily newspapers and weekly magazines,
in the building of weekend chalets and condominiums, in new
154

ski lifts and expanded restaurants, in the increasing concern for 'overloading' the environment, and in the interference of 'those Parisians' in *our* affairs. A national periphery is no longer as tranquil and peripheral as it once was. And these notions of centres and peripheries can be hierarchically arranged, so that a mountain village in France, like Autrans, may be on the periphery of its regional centre Grenoble; Grenoble may be peripheral to the nation's capital Paris, and Paris . . . ? Well, a nation's capital may be peripheral with respect to a still larger centre, or it may, like Paris, be close to the centre of an even bigger geographic entity like Europe. So what about Europe as a whole? Since we are really talking about the potential for people to interact with one another, it should come as no surprise that this macro view of human contacts in space and time should be the work of Gunnar Törnqvist, who we met in Chapter 8, a colleague of Torsten Hägerstrand at Lund (Chapter 12).

When we think about individual contacts on these scales, we must remember that 80 per cent of air traffic consists of people conducting business, both private and governmental, despite all the tourist flows to sand and sea in the summer, and to snow in the winter. For all the telephone contacts and conference calls, there is still nothing like sitting down with people on a face-to-face basis for that sort of frank and full communication that leads to explanations. understanding and agreements. So a person's ability to contact others on a face-to-face basis becomes a very important matter in the world of business and government – not to say in the world of scholarship, where the main commodity is ideas.

Suppose we take 98 major urban centres in Europe, and ask this question: if you do not leave before 6 in the morning, and you want to be back before midnight, how many hours of contact can you have with the rest of the European 'system' (Figure 13.10)? Mapping the values relative to Paris (100), we see that old core of interaction potential appear once again with a spillover to London and southeastern England. Paris, Antwerp, Brussels, Amsterdam, Rotterdam, Lyon, London, Geneva, Berne, Hamburg, Zürich, Düsseldorf . . . I am not sure how you would calculate power and wealth exactly, but there it is, right at the centre. Another ring of secondary contacts (and power?) extends from Helsinki (41) to Madrid (39), and beyond that the peripheral towns of Lisbon (14), Athens (17), and Reykjavik (1).

Quite tragically, there is one further feature of the map that tells us that *potential* interaction may not always be realized.

Figure 13.10: The European 'stay time' map of potential face-to-face contact if someone starts a journey at 0600 and returns by 2400 the same day. Values are mapped relative to Paris (100) the peak on the surface.

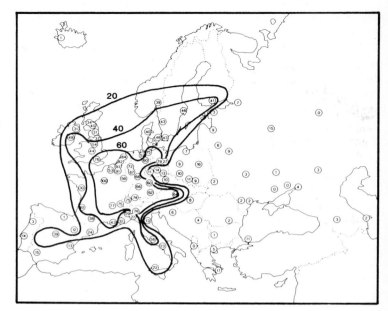

With the exception of West Berlin (79), there is not a single value greater than 16 (Warsaw) in eastern Europe, a direct and empirical cartographic statement of the way political constraints can reduce potential face-to-face contacts, reduce explanations, reduce understanding and reduce agreements. What we see here is the human agony of the Iron Curtain. For those living on both sides it is very real, and the raw manifestations of the constraints are humanly appalling. I recall a trip by hydrofoil along the Danube from Vienna to Budapest, passing the Czech border with its high multiple fences, watch towers with guards peering through binoculars, and machine guns tracking the boat through the frontier. No one was allowed on deck, except to go to the toilet two steps away, where a large and watchful guard made sure you went no further. Or, another time, stopping for an hour at Moscow airport, from Tokyo to Paris, and being met by guards with machine guns at the foot of the ramp, more on the bus, still more at the foot of the steps leading to the isolated transit lounge, and yet more at either side of a short balcony where you could stretch your legs after a long flight. With all that suspicion and fear, no wonder potential measures of human contacts and interaction are warped by political reality.

And because of this reality, it sometimes seems that it is not considered politic to write about these things. We do not want to upset 'them', because we might not get visas the next time

156

we apply. Sometimes even professors do not swear to tell the whole truth these days, only that which is politically advisable. But there it is on the map, a human constraint on a human space, a Europe split in two. It is not a new division, and this is not the first time that geographers have looked at such differences from a macroscopic point of view. At the end of the nineteenth century, Halford Mackinder, who founded the first department of geography at Oxford University, was writing about the earth divided into a World Island with its Pivot Area or Heartland, surrounded by Marginal and Outer Crescents on the Periphery (Figure 13.11). When you look at it like this, on

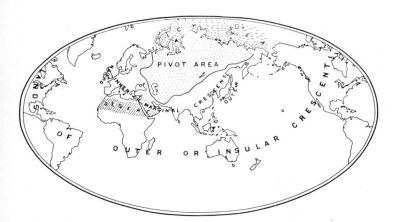

Figure 13.11: The core (Heartland) and periphery (Insular Crescent) map of Halford Mackinder, drawn in 1919 on a flat map.

a flat map, you begin to understand such geopolitical terms as 'containment', and we have all seen maps in weekly magazines of a Soviet Union surrounded by menacing bases. Since she is surrounded on all sides, 'no wonder' she tries to place a ring of buffer states between herself and that hostile 'outer world'. No wonder – unless you happen to be a buffer state.

Quite apart from the fact that the presence of buffer states discloses the two-dimensional thinking of a bygone age, thinking that hardly seems appropriate in this three-dimensional world in which every man, woman and child can be reached by atomic missiles in 20 minutes, it also points to the way our thinking is conditioned by two-dimensional maps. If we are really prepared to think globally, and to stretch and shrink space, then perhaps the political earth is really like a balloon (Figure 13.12). Hold it by Mackinder's Heartland, and the periphery puffs up menacingly, but hold it by the Periphery and the Heartland looms large and dangerous. In either case, the poor buffer zone remains, for this is what a mathematician would call a topological transformation. Things, like our

Halford Mackinder
University of Oxford
1861–1947

157

Figure 13.12:
Mackinder's Heartland
and Insular Crescent on
a balloon world.
Depending upon where
you squeeze, you get
the view you want to
see.

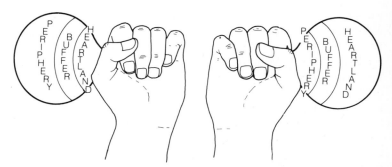

political world, can get out of shape, but for all the stretching and squeezing the *connections* remain the same.

Our macro-geographic viewpoint has led us from pressure surfaces, income fronts, winds of influence and patterns of contacts to political divisions too often enforced by military might. Here we have touched upon geopolitics in the raw, and geographers have long had things to say about the geopolitical state of the world. This has brought them into contact with the military of all nations, points of contact perhaps where the pens of scholars cross the swords of soldiers. But these swords often have two edges, and we really ought to examine both of them.

Part IV
Three double-edged swords

Geography and the military

14

The division between centres and peripheries in geographic space can direct thinking to many aspects of our human condition – social, economic, political and, perhaps as an ultimate aspect of the latter, the military. As we saw with Mackinder, and his notion of a great land mass of a Heartland, geographic space may well take on a high strategic value, and the idea of trading space for time has been a familiar story to every Russian from the time of Napoleon to the scorched earth policy of the Second World War in the face of the Nazi invasion. Of course these ideas, relating almost entirely to land warfare, have been greatly modified over the years, starting with Mahan's monumental appraisal at the end of the nineteenth century of the strategic value of sea power. His ideas have been slowly modified in their turn, perhaps a bit more rapidly since that day in 1923 when Billy Mitchell sank an old battleship off the North Carolina coast with an airplane. This proved to be yet another turning point in warfare, for now the third dimension came into strategic play with a vengeance, a lesson bitterly learnt at Pearl Harbour on 7 December 1941, and by the loss of the cruisers *Prince of Wales* and *Repulse* off Malaysia three days later. In the 1980s, the battles off the Falklands have demonstrated that aircraft at supersonic speeds can overwhelm almost any counter-system, for air-launched missiles can appear from below the horizon, skimming the surface of the sea to make radar detection extremely difficult. On a world scale, that third dimension of the air allows intercontinental ballistic missiles to penetrate *anywhere*, and as the geographer William Bunge has noted, this is like staking out every man, woman and child on the front *plane* – because if one-dimensional front *lines* are boundaries and frontiers in two-dimensional space, then two-dimensional planes are their equivalent in three.

It seems to me that geography (using the word rather broadly), must enter into military strategy – it always has, and presumably always will. One of the basic reasons, of course is that waging war effectively requires you to know where you are, where the enemy is, and what it is like in between. For the past 300 or 400 years that has meant using maps, and the more accurate and detailed the better. So it comes as no surprise to realize that the armed forces of a nation have often been the organizations involved in the compilation, drawing and production of maps and charts at all scales, and even until a few years ago some of these were kept under lock and key – an exercise in futility today, when high resolution air photographs and satellite images are available on a daily basis for almost any portion of the globe. Still, old habits of secrecy die hard, even if you do happen to photograph me today from 24,000 metres up just as I am locking up my classified maps made in 1923. Most European countries had their basic topographical sheets compiled originally by their armies, and the Ordnance Survey of Britain still retains that adjective *ordnance* that the dictionary says means 'mounted guns and cannon'. At the end of the twentieth century, most armies have sophisticated research units doing 'terrain analysis', studies of the load-bearing capacities of soils under various conditions of temperature, moisture, slope and so on. Even in these frightening days of atomic warfare, terrain can still be an all-important factor in waging war. It can provide almost total cover, as Fidel Castro and Che Guevara knew in the Sierra Maestra of Cuba, and as the bicycle porters knew on the Ho Chi Minh trail of Vietnam. As for the Pathans of Afghanistan, they have always known the value of the high ground – from the arrogant nineteenth-century British expeditions 'To teach those chaps a lesson', to the recent invasions by a modern Russian army equipped with tanks, helicopters and missiles. Teaching the Pathans lessons in warfare on their own ground has always had a tendency to backfire.

These days, remote sensing from satellites (Chapter 18) provides direct coverage of the earth's surface, and virtually all the research that has gone into improving these images has been inspired by military considerations. Indeed, the resolution of the instruments used for military intelligence is so extraordinarily fine that satellites for civilian use (LANDSAT I launched in 1972, and LANDSAT V in 1984) have had their imagery deliberately degraded. Technical know-how is obviously of military value these days, and remote sensing is a particularly sensitive area, although it sometimes seems that security classifications are jacked up to quite excessive levels. A few years ago, a large professional organization in the United

States held an annual conference on the eastern seaboard, and wanted to display a large satellite picture of that part of the country. Sorry . . . too sensitive . . . security forbids the publication of such an image at the present time. So the organizers approached the Russian Embassy in Washington, whose personnel were simply delighted to donate a fine picture taken 'just the other day'.

By the very fact that wars take place on or around the surface of the earth, and the earth has always been the proper province of the geographer, so geographers and geography are going to be involved in war. It seems to me that serving one's country in the time of a just war with one's professional skills is meet and right, and geographers have much to offer, particularly in the area of military intelligence and strategic planning. During the Second World War, for example, a number of American geographers served with distinction in the Office of Strategic Services (OSS), the wartime organization of 'Wild Bill' Donovan that was to become the Central Intelligence Agency (CIA). Waging war on a global scale – which was precisely the scale of the Second World War – often requires detailed knowledge of a geographic nature, and much of this information is available on maps, or is most effectively presented by way of maps. Knowledge of the physical terrain, the strategic resources and the way a potential region of battle is organized by roads and communications is absolutely vital – and I mean *vital* in the literal sense that this constitutes information that can save the lives of your soldiers. As a former soldier, this has always appealed to me. Geographers have also contributed to wartime efforts on the home front. In Britain, for example, Dudley Stamp conducted a detailed land use survey of the island in order to evaluate the agricultural potential of a nation that was fighting for its life. Every ton of food that could be grown at home meant an extra ton saved on ships passing through the perilous waters of the North Atlantic. It was a quiet and unspectacular effort, but it contributed to Britain's defensive capabilities at a crucial time in her history.

However, it is not simply in wartime that geographers may wish to give of their skills in what they believe is a just cause. I know of a number of geography students who have gone into defence mapping agencies and intelligence operations, and sometimes someone with a combined military and geographic background can illuminate some of the deep human problems that lie at the heart of many potential conflicts. Brigadier-General Richard Eaton, for example, has conducted careful interviews with staff officers of various nations, and created

Dudley Stamp
London School of Economics and Political Science
1898–1966

163

Three double-edged swords

dramatic images of what we might call the perceived military geography of these men – the images they hold of friendship and threat, size and location, strategic posture and the bases of power. For example, the strategic image of a Venezuelan staff officer generates a rather hostile world (Figure 14.1). The

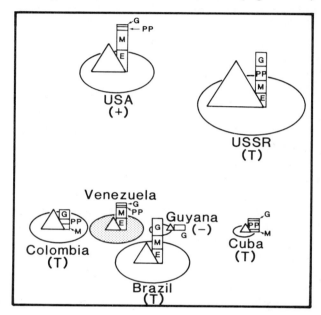

USA and USSR are both prominent on this 'mental map', with the power threat of the latter recorded mainly on what was termed a psycho-political scale. Even tiny Guyana next door, with its minute armed forces, appears to be regarded as a threat to Venezuelan national integrity. The officer also greatly underestimated the size and population of Brazil, and evaluated the threat of Cuba almost entirely on psycho-political grounds. In brief, here are the skills of a modern geographer brought to bear on the subject of potential conflict through the mental images of those trained to wage war. Such images point in very concrete ways to problems of threat perception, problems that should make us all much more sensitive and aware of the role of images in strategic decisions.

But . . . and it is a big *but*, the engagement of geographers in military intelligence, planning and research can indeed be a two-edged sword – the first of three we are going to consider in this section. Now to say this at one level is trite – nearly all knowledge can be used for good or ill, so there is nothing particularly new here, even if many people today forget this

164

truism. But at another level, the involvement of geographers with the military has caused grave concern, and in the 1970s many voices at professional meetings were raised expressing a deep sense of disquiet. Part of the disquiet was simply one aspect of a larger, overall concern that many were feeling (and continue to feel), the sort of deep concern expressed by President Eisenhower, one of America's few 5-star generals, in his last address to the nation before he left office – an address warning of the grave dangers to the democratic process if the 'Military-Industrial Complex' continued to strengthen its hold. In today's confrontational world, billions of dollars are spent each year on armaments, as the Stockholm International Peace Research Institute in Sweden documents on a regular annual basis. These amounts of money mean huge profits to firms of all nations producing and selling arms, and the temptations are clearly immense, and often not resisted with much enthusiasm by those out to make a large pile of money. Moreover, a title of a geography article such as 'Military payroll and regional growth' discloses the obvious fact that spending on arms and other military requirements means huge flows of money to certain places, with all the new jobs, high employment levels and extra business profits for the region. In the United States, southern senators tend to be very senior, on most of the important committees, very powerful as a result, and not averse to steering a disproportionate amount of military spending to their own states. After all, that is one of the major reasons their constituents voted for them. A number of careful geographic studies document the impact of military spending on a region, and highlight the way the local economy would collapse if a large military base or factory were to close suddenly.

Another part of the disquiet over geography's involvement with the military arises out of a deep concern for what this does to geography itself, to its directions and motivations for research – in brief, to its scholarly integrity. Some feel that no geographer, and no professional geographic organization, should have the slightest thing to do with any aspect of the military, and that all research funding from any military source should be refused. Sometimes the voices and the writings are quite shrill. In France, Yves Lacoste wrote his book *La Géographie qui sert, d'abord, à faire la guerre (Geography Which Serves Above All to Wage War)*, a slim polemic that lashed out at technology, government bureaucracies, urban, regional and national planners, the intellectual distortions in geography caused by the quantitative revolution, similar distortions caused

165

by military involvement . . . just about everything you can think of. And yet, as in many polemical statements, there is underneath the tirade *a* truth, and few can fault Lacoste for the depth of his concern. On the other side of the Atlantic River, William Bunge has devoted much of his life to exposing social injustice and the horrors of atomic war, using the same polemical stance in an effort to grab people's attention and direct both their humane and professional thinking towards these worldwide problems.

There is no question that the two-edged sword is there, but to label geography a discipline whose major purpose is to wage war is, to my mind, quite absurd. To condemn geography because a few geographers lend their skills to military intelligence is like condemning mathematics because important research areas of combinatorics are at the heart of modern cryptography and intelligence operations. For all practical purposes, unbreakable codes are now possible because very deep theorems in number theory, originally investigated out of pure intellectual concern, underpin coding procedures today. So is mathematics a subject that is to be characterized, above anything else, as one that wages war? I think reasonable people would disagree.

Nevertheless, the tension is there, and ultimately it must be left in a free society to individual decision. My own feeling is that both the quietly expressed concern and the polemical tirades have sensitized all geographers to the dangers of too close involvement, and have pointed to the way such 'cosying up' can shape and guide the intellectual life of a discipline. Large amounts of research funds from military sources are always tempting, perhaps all the more so because people in the academic world tend to be underpaid – as teachers are everywhere. But the fact is that research funds from *any* source always constitute both a benefit and a danger. Money is power, and all power corrupts – eventually. Too simple an aphorism, of course, but the problem of influencing the direction of geographic research is there, and the temptation to exercise power in order to steer it in certain directions and not others is always very great. Particularly if you think you are at the centre of things and have *the* truth. So this question of managing geographic research is an important and integral part of modern geography, and it is something we should take a closer look at.

Managing geographic research

15

The second of our two-edged swords opens up another of those tension-ridden areas in which there are both benefits to be gained and dangers to avoid. Providing money for geographic research ultimately comes down to the question of *managing* geographic research, because decisions to support, or not to support, particular lines of inquiry often mean that a research project will go ahead, or be turned completely aside. Providing research funds is rather like turning a financial tap on or off, and those who are involved in such decisions bear a heavy responsibility to their offices, to the applicants and to future generations of scholars. These problems are by no means unique to geography, but characterize all funding decisions across all the humanities and the sciences. We have to evaluate the worth of a proposed idea now, and try to determine what its impact will be for a particular area of inquiry and understanding years later. This is no easy task.

However, it is a comparatively recent task as far as geographers (and, indeed, many scientists) are concerned. Before the Second World War, there was virtually no government funding of geographic, or any other sort of scientific, research – that is, funding that ultimately came from the taxpayer's pocket, or from the investment portfolios of large charitable foundations excused from paying taxes. Most universities could make a few, rather meagre contributions to individual scholars, perhaps from special bequests and endowments set up for particular disciplines. These might pay the overseas travel expenses for fieldwork, but little more. The great foundations of the United States and Europe (Ford, Rockefeller, Guggenheim, Nuffield, Volkswagen, German Marshall, Werner-Gren, etc.) were only starting in, or still on the horizon of the post-war age. Geography, like many disciplines in the university, jogged along with a rather low research profile.

All that changed with the Second World War, particularly in the United States with its high degree of involvement of geographers with the war effort. The general public might go on thinking that geography was knowing the capital of North Dakota, but those responsible for military strategic decisions and intelligence knew much better. Some of the first support for geographic research came in 1955, when the Office of Naval Research started a modest programme of overseas fieldwork and research through the National Academy of Sciences. In addition, the Ford Foundation started its Foreign Area Fellowships for young doctoral students wishing to conduct their dissertation research abroad. These programmes had a large impact, because for geographers it is particularly important to get outside of their own culture and country to examine their models and presuppositions under very different conditions. It was also in these years that the National Science Foundation started its geography programme, and it provided extremely valuable support for short conferences on the new ideas, methods and concepts, as well as helping to develop summer workshops that allowed the new ideas to reach many more people directly, and so spread more quickly. External funding for research and training rose from $438,000 in 1962 to $595,300 in 1964, and today (the mid-80s) it averages around $10–12 million per year from all sources.

Now when sums of money like these are involved, someone, somewhere, has to manage them. Faced with a pile of applications exceeding the funds available, decisions have to be made about who gets what. Generally speaking, applications for research funds are reviewed by 'knowledgeable people in the field' who remain anonymous to the applicants, although in the United States their written evaluations are now made available as a matter of course under the Freedom of Information Act. In contrast (and unlike many review procedures for assessing submissions to professional journals), the applicants are not anonymous, and this has given rise to the so-called 'halo-effect', where the prestige of well-known names and their university affiliations enhance the evaluations perhaps more than the strictly scientific merits of the research proposal warrant. In a series of experiments conducted in another field of the human sciences, highly regarded papers by well-known people had the names and the university affiliations changed – either up the hierarchy of academic prestige, or down the ladder to Flotsam and Jetsam University. Reviewers tended to give adverse evaluations to articles from ol' F & J, and accepted articles from prestigious universities that had previously been turned down by a whole series of reviewers. The same halo-effect will affect *any* evaluative procedures,

168

including those lying behind decisions to fund research, unless the anonymity cuts both ways.

Unfortunately, even if anonymity were to cut both ways, there are few research-oriented geographers who have not had the experience of receiving an 'anonymous' article or proposal, and then making an almost certain guess about who the author really is. The footnotes are often a dead giveaway. The ethical thing might be to return the proposal, or indicate that a 'good guess' had been made about the author. But I suspect that few do this. It is an imperfect human world, and in the middle of supposed very cool, calm and objective scientific evaluations, where nothing but the pure merits of the proposal are considered, what we actually see, time and again are very highly charged human emotions. Every discipline is split into feuding camps of one sort or another, and proposals compatible with the norms of one camp will be lavishly praised and recommended, while if they fall into the hands of reviewers from the opposite camp they will be condemned. Those who decide who will do the reviewing exercise enormous power – and, once again, all power corrupts. One programme was notorious for years because it gave funds only to members of a small but powerful 'school', a senior coterie whose members did their utmost to perpetuate their particular views, and used the financial leverage in a particular funding area to do precisely that. Young geographers, particularly those associated with the 'quantitative revolution', found it impossible to break into the closed circle because of its self-perpetuating nature.

It is here that we come to the heart of the research management problem – the self-perpetuating nature of many of the committees and advisory boards that actually end up managing large sums of money, so affecting directly the research directions of a discipline. We must not exaggerate here: it is not a matter of a few masterminds pulling the strings of some dancing marionettes, but rather quite a number of people doing their best, but nevertheless giving some fairly hefty collective shoves to a large and heavy elephant. Still, with enough shoves in one direction and none in another, even an elephant is likely to lumber off in that direction. The problem of an entrenched 'establishment' is always a difficult problem, no matter what the discipline, and it is distressing to observe how the Young Turks of one generation can so quickly turn into the Old Guard of another. In one well-known institution of the social sciences, new members are chosen and confirmed by those already on the guiding committee, people who were selected by the same process. Despite a scattering of prominent names, most are not noted for their dynamic imagination and creative *avant garde* thinking. I have personally witnessed serious, straight-faced discussions about Dr X and Dr Y, which noted that while Y is a much more imaginative scholar with considerable research experience, X was 'a better committee man' – a verbatim quotation, and one that many will approve. It was run-of-the-mill X, who was never known to rock the boat, who got the appointment, so perpetuating the nice and safe view of things.

But the safe view of things in a science is what the historian of science Thomas Kuhn has called 'normal science', the sort of generally worthwhile, but often not terribly exciting, science that plods away doing things along the usual lines of the day – according to what has been termed the accepted *paradigm*. By definition, normal science is establishment science is acceptable science is don't-rock-the-boat science, and this turns out to be the most easily managed science. Proposals that are approved are unlikely to be terribly controversial, although, equally, they are unlikely to lead to anything particularly startling or new. I think it is fair to say that most really far-reaching ideas seldom come from officially sanctioned and funded research. After all, who would give a clerk in a patent office in Zürich any more for those madcap ideas about space-time? Damned fellow wants to use a completely different area of mathematics called tensors, and who understands these? The clerk's name was Einstein. As for that bunch of people in Copenhagen, they can never make up their minds about anything – different model of the atom every week, and *none* of them work. No use throwing money away on them. So Niels Bohr would be turned down too. As for that fellow Galileo, tell him not to bother to forward his application! Actually, this is a terribly unfair parody on research in physics, unfair because one of the quite extraordinary things about certain areas of modern physics is the sheer quality of its imaginative thinking. It almost appears today that doing frontier research in physics requires thinking the unthinkable, postulating ideas that are contrary to all experience and accepted views, and being willing to stand open to new and quite strange possibilities. Although they have also had long periods of 'normal science', the physical sciences generally have a strong record of supporting new ideas. In contrast, people in the human sciences tend much more to 'play it safe', and their cautious, conservative attitude is predicted in part on the nature of their subject. At official bureaucratic levels, such caution turns to fear when people on the staff of a man like Senator Proxmire of Wisconsin pick out seemingly strange titles of research projects for his Golden Fleece Award.

This was a pseudo-award 'given' to research projects which the senator thought were too esoteric or trivial, and so their funding from the public purse 'fleeced' the taxpayer who supported them. Since being sued by an anthropologist, and deciding to settle the damages out of court, Senator Proxmire has tended to make such awards less frequently.

Other problems also arise, of the sort generated by any bureaucracy. Secrecy is probably the worst, and it can work in many ways. Firstly, there may be, for example, an unspoken and unrecorded decision to spread funds over many small projects, rather than concentrating them in fewer, but scientifically much more competent, hands. Spreading the money around is always politically popular, and those who decide on such a course will be rewarded by the sight of many eyes welling with the tears of gratefulness. On the other hand, and in terms of providing adequate support for those who can undertake genuine research, such a policy is likely to fritter away and waste enormous amounts of money. Secondly, the secrecy of the review process allows reviewers to hide behind the skirts of a major funding agency, and never be held responsible for their criticisms. In new, and almost by definition controversial areas, criticism should be published (in the literal sense of 'making public'), with the names of the critics on the masthead. This is how an open science has always proceeded in a responsible way. But reviewers and critics who know they will never be held to account for their often sharply worded criticisms leave a scientist feeling angry and impotent – emotions that are not conducive to imaginative and illuminating thinking.

A further problem arises when those with official positions in funding agencies, with considerable power to influence the allocation of research funds, begin to publish personal views about what they consider to be desirable research directions in a discipline. At this point, they are wearing two hats – that of the bureaucrat, and that of the research scientist – and the roles are totally incompatible, despite the usual *caveats* that 'these views reflect the opinion of the author, and do not constitute the policy of the funding institution with which the author is associated' – or words to that effect. Such statements are essentially nonsense, and constitute little more than legal mouthings. Unless such a person is totally schizophrenic, it is impossible to partition and compartmentalize the private views from the public policy when the same person is involved in both. The problem is compounded when errors and misinterpretations in a field are taken over as personal viewpoints, and then elevated to the level of semi-official policy. These are

171

particularly difficult to correct, and they can shape the future directions of a field by dissuading many from submitting proposals that try to correct the errors, or perhaps advance knowledge in other, often quite new directions.

As for other sources of research funding from governmental agencies, they tend to be highly task-oriented, meaning that they usually want a specific job done in a specific time, and almost invariably along well-known and familiar lines. Approaching difficult and creative work with a scheduled mentality, 'the rules' often require so many monthly and interim reports that there is little time to do any real research. It is yet another example of those contradictory, two-edged swords: many in government would accept that in complex areas of engineering, say marine, aeronautical or atomic, there might be at least ten years between an initial idea and the actual production of a working model. Yet in areas of much greater complexity of the human sciences, the same mentality seems to expect research in the human realm to go from an abstract model on paper to full-fledged application in a matter of months.

Finally, geographic research, like any other human enterprise, is shaped (and is to that degree 'managed') by the values of some who have the power to provide or withhold information. Science is, or should be, an open human endeavour (subject always to strict rules protecting the human beings who may be the subject of inquiry), and that means an openness to inquiry, and an openness to the dissemination of the findings. Thwarting the free flow of information has no room in such enterprises. Unfortunately, some geographers have come up against institutional and personal values that make the dissemination of their work additionally difficult. Richard Symanski, for example, suffered grievously for his important work on prostitution, and his *Immoral Landscape: Female Prostitution in Western Society*, published by Butterworths of impeccable scientific credentials and reputation, was denied institutional mailing lists to send standard book announcements on the grounds that 'a number of our members would be offended by our participation in the promotion of this publication'. He was obliged to fight a long and vituperative battle, and finally managed to get the initial decision reversed, but after the damage had been done.

So here is another contradictory aspect of the story of modern geography. On the one hand, the availability of substantial research funds in geography has done much to support the dramatic changes over the past couple of decades,

Richard Symanski
Independent Scholar
1941–

172

and much of excellence that has been accomplished could never have taken place at the same pace without such important financial backing. Agencies such as the National Science Foundation in the United States, the Canada Council, the Social Science Research Council in Britain, the Centre National de la Recherche Scientifique in France, and other similar national and private sources have done much to push forward the very exciting changes and altered perspectives. But there is equally no question that some price has been paid, to the degree that power is irresponsibly exercised, either by exerting some control over certain lines and directions of research, or by allowing criticism to hide behind the large feather pillows that all official bureaucracies ultimately become. These are surely the great curse of the modern era, no matter what the political persuasion and stance of a particular country, and everyone feels a deep sense of impotence facing them precisely because it is impossible to pin down responsibility for a decision. It is always 'policy indicates . . .', where policy is often unstated; or 'the rules dictate . . .' when the rules were written by a faceless committee now long disbanded; or 'the former chairman's decision . . .' when the former chairman has long done a Pontius Pilate act. Scientific funding via bureaucracy is no different, and constitutes simply a special case of a much larger social problem. It generally supports the solid, 'normal', safe if somewhat unimaginative and unadventurous science, and seldom dares to set aside funds as 'risk capital'. This means that few really original breakthroughs in thinking are made from sponsored research, although support may come later once the initial, and often most difficult running has been made. There is a strong conservative streak in most of the human sciences, not the least because new ideas always challenge the adequacy of the old, or demonstrate how terribly constrained the old ideas actually are. Pointing to the constrained and inadequate nature of current thinking may not be terribly popular with many of those who have invested a great deal of time and emotional energy gaining an intellectual command of the *status quo*, and who now form a powerful part of the Establishment. Geography, as any scientific and human endeavour, is full of human imperfections, but openness and full disclosure seem the best ways to improve it.

While considerable research funds go to theoretical work, it is frequently considered politic and wise to support projects that can demonstrate practical applications of a particular line of research, or how an approach can solve real problems – for many people, the adjective 'real' has a sardonic ring to it, for the 'real' problems may be theoretical after all! Nevertheless, useful applications are clearly important, and applied research is a crucial component of the overall picture. Much of it is conducted through consulting, when the particular expertise and skills of a geographer are hired by a governmental agency, a commercial firm, an international institute, or perhaps a private charitable foundation to advise on the direction of a particular decision, or on the solution of a difficult, often technical, problem. Many geographers who do consulting find the experience and challenge immensely rewarding, either as full-time occupations, or as occasional forays from their normal life of university research and teaching. Often teaching is greatly enhanced by the 'case studies' and first-hand experiences that result from consulting activities, and in many areas of geography students seem to have more confidence in teachers who sometimes practise what they preach. However, consulting is also an area that is contradictory and tension-ridden, with pros and cons, benefits and dangers. It is time we looked at the third of our two-edged swords.

The Beltway Bandits 16

Outside of what we might call 'traditional engineering circles', the consulting firm was almost unknown before the Second World War. By a consulting firm I mean some sort of organization that is willing to make formal bids for contract research, staffed by people who have the skills to carry out investigations for clients who come to them with particular problems to be solved. These firms today are both large (some very large indeed, with gross receipts over $100,000,000 per year), and small (sometimes just a single individual), and the problems they deal with are also extremely varied. They do research and write consulting reports on structuring and managing organizations, the social impact of proposed economic policies, environmental impacts, the cement needs of a country, large lumber mill locations and the local ecology, architectural design in the Arctic, pollution control, jellyfish and the tourist trade, town, city and regional planning, natural gas demands and pipelines, influenza epidemics, new agricultural schemes, sewage disposal and water management . . . the list is almost endless. It is all a reflection of a growing awareness that we just cannot go on in a hit or miss fashion, but that some planning, some thinking about consequences in the future, is necessary to create a decent and humane world in some sort of balance with Nature. Notice that many of these topics – and literally hundreds more I could cite – have a very geographic 'feel' to them as they concern some intimate connections between human beings and the earth they live on. Indeed, one of the reasons geography has exploded to such prominence in the last thirty years is because people are becoming much more aware of the spatial dimensions of their societies. How geographic space is organized, structured, treated and reorganized over time is seen as an increasingly important issue in many societies around the world.

It is to deal with many of these problems (most of which take considerable amounts of information and careful analysis to clarify, let alone solve), that hundreds of consulting firms have sprung up, and in the United States many of them align themselves along the big ring or beltway roads that are such a characteristic of America's cities – Route 128 around Boston, Route 695 around Baltimore, and perhaps especially Route 495 around Washington, DC. It is for this reason that they are sometimes called the Beltway Bandits – a term used in a somewhat tongue-in-cheek, self-depreciating sense, and not necessarily in a pejorative way. In the United States, many consultants are to be found around Washington precisely because they can back up many of their written proposals and bids with personal contacts, for the Federal Government alone is a huge source of funding for environmental, social, economic and military research. But a location in a nation's capital is typical of these firms: London, Paris, Tokyo, Rio . . . they all have their share of them. It is not unusual for these organizations to call in people with particular expertise in certain areas, and many of these external consultants are from the universities. In fact, virtually all of the personnel of consulting firms are from the universities, particularly postgraduate students, or former faculty who found they could increase their salaries enormously – sometimes by an order of magnitude or more. Concretely, that means an associate professor at $25,000 could make $250,000 and on up after a few years running his or her own firm. Full-time consulting can be very lucrative if you are good at it and can build up a solid reputation for delivering the goods.

Given the increasing awareness of problems that time and again are essentially geographic, it is quite natural that those with a contemporary geographic background find themselves being snapped up by consulting firms, government agencies and research institutes. As I noted before, I myself have had many former students going into consulting work, some of them rising to positions of considerable responsibility in the World Bank, the Agency for International Development and numerous private firms. Some, after gaining experience, have set up on their own, and their activities take them around the world – like any good geographer. Without exception, they have made technical skills an important part of their academic programmes, and nearly all have gained considerable overseas experience, and usually have competent linguistic skills as well.

What sorts of problems do geographers tackle when they undertake consulting? As we saw above, the range is extremely varied, but let me pick out just a few examples. One major figure in contemporary geography, Brian Berry (Chapter 9), has devoted much of his professional life to problems of urbanization, in his adopted country of the United States as well as in the Third World. The problems are very different: in the USA, the 'hollow doughnuts' are growing as many of the centres of America's cities lose their old employment and tax bases. Some of the old decayed cores are in a chronic mess, and few know what to do. In contrast, the Third World cities are exploding as ring after ring of shanty housing surrounds the older core areas. In Indonesia, for example, people are still streaming into the main cities, despite attempts by the government to open new agricultural areas for them, and to provide advice and guidance for urban planning in these desperate situations takes careful and experienced analysis. Because of their more holistic, less limited viewpoints, and because they so often insist upon gathering and dealing with the concrete facts in their geographic, usually mapped, relationships, geographers are increasingly prominent in such areas of integrated planning. This means thinking through all the interconnections and implications of new housing, roads, water supplies, basic amenities, sewage . . . all the things that are so tightly inter-related in a modern expanding city.

Of course, not all geographic consulting goes on in the cities. Geographers were prominent in helping Afghanistan formulate its first national census in the 1970s, working closely with the people to think through what information was most useful for a poor country trying to plan the best use of its human and natural resources at the national and regional levels. In many Third World countries (Chapter 23), rural areas have also received considerable attention, particularly helping people to think through the stage-by-stage development of small feeder roads, what the order of priority should be, and what the likely impact of a new road might be one, two or three years after completion. Geographers are acutely aware that the impact of a road is by no means confined to the economic level: in many ways the social impact may be even more profound as new ideas and innovations move more quickly, and people move more quickly too – frequently *out*! Along with improved roads come all the other possibilities: new schools, new rural infirmaries, new birth control clinics . . . new possibilities of all sorts. We know now that these are the things we call central place functions, and a number of geographers have contributed greatly to the thoughtful provision of better amenities in developing regions, or in older established ones. Gerard Rushton, for example, has directed much research as a consultant on the provision of medical care, both in India and in his home state of Iowa as the director of a major research institute, and we shall look at his work in more detail later (Chapter 23).

Working as a consultant on transportation development seems to provide a particularly good vantage point on the entire development process, because when you focus on the 'connective tissue' of a region or country you become extremely sensitive to other 'interconnections' between land and people, city and countryside, agriculture and forestry and mining, all of whose products have to move on the roads or rails. A number of geographers who started such consulting in the 1960s have now moved to prominent international positions, not the least because their integrative viewpoint is becoming increasingly valuable in a world of marked specialization.

Such an integrative perspective is valuable right down to the individual level, and an older tradition of careful fieldwork is becoming increasingly important, particularly when it is informed by new possibilities and approaches for structuring and analysing large amounts of detailed data. One of the distressing things about making recommendations to small farmers around the world is that all those new seeds and techniques and marketing methods that look so exciting and well-proven on paper, or seem to do so well in the finely groomed demonstration plots, do not seem to be appreciated or arouse the same enthusiasms in the farmers themselves. Too often the poor farmers are labelled 'conservative' or 'ignorant' by the visiting expert (?), when they may really have very good reasons for sticking with the old ways. Graham Chapman, a geographer with considerable consulting experience in dryland and irrigation farming in India and Bangladesh, has conducted detailed analyses of individual farmers and their farm operations, and has come away with an increased respect for the astute evaluations and decisions they make within their own worlds. They often seem intuitively aware of all sorts of connections and ramifications between elements of the natural world, and the human world of crops, implements and social relations, that the 'expert consultant' is unaware of. Chapman has incorporated much of his field-derived insight into an important tool for teaching, and we shall take a closer look at his 'Green Revolution' later on (Chapter 22).

A good consultant is always sensitive to a particular cultural situation, although there are equally occasions when a consultant 'from the outside' can perform a very valuable service by being tough-minded. At a higher institutional level, this is exactly the role the International Monetary Fund plays, virtually forcing financial reforms on governments who are often secretly grateful since they do not have the political courage to take such steps themselves. But whether at the international or local level, consultants have to find that delicate balance between sensitivity to local conditions and direct, look-you-straight-in-the-eye advice. Almost by definition, the problems are complex (or consultants would probably not have been called in), and often there is not a clear and unequivocal 'best' answer. As is true in many areas of life, you cannot please everyone at the same time.

It is not that geographers and others are uncaring in their roles as consultants, and give advice in an unthoughtful fashion. All the ones I know take their consulting work and responsibilities very seriously, and give the best advice they can. Occasionally you will hear a horror story, but they are becoming rarer today. I remember in the late 1950s hearing about one consultant (not a geographer!), who recommended two enormous bulldozers to dig small earth dams in northern Ghana to catch the fleeting seasonal rains, and so extend the crop season considerably. After all, this is the way we do it back home in Utah, and what's good enough for Utah . . . but you get the idea. The problem was that 200 daily labourers digging and carrying earth on their heads were laid off, diesel fuel and the wages of skilled mechanics and operators cost a packet, and when the machines broke down spare parts had to be trucked in or flown up from the coast to the inaccessible interior, often with delays of up to six weeks because the orders had to go right back to the factory in the USA. When the rains came the dams were not ready, although they could have been completed in time, and at half the cost, using the traditional way.

But . . . and it is another of those big *but*s: that tongue-in-cheek term, the Beltway Bandits, sometimes has an epithetical ring to it as well, and here come the tensions and contradictions of the third of our two-edged swords. There is no question that consulting raises grave ethical questions, and few who have undertaken it have not felt such tensions at some point or another. I know of large consulting firms who have advised on planning African cities whose analysts had never set foot on African soil, let alone in any of the cities concerned. And sometimes recommendations are made by international consultants which are so inappropriate that one wonders how they could have been put forward by rational people. In Egypt, a new city is growing north of Cairo, and the first plans recommended that traditional, and highly effective, methods of garbage collection and disposal be replaced by huge machines

179

crushing everything down under tremendous pressures into solid cubes. In this part of the world, Coptic Christians traditionally collect and sort the garbage materials, raising pigs (a valuable source of protein) on the waste food and vegetable matter, sending 18–20 tons of waste cotton cloth per week abroad for fine paper making, and salvaging and recycling everything down to the carbon rods of flashlight batteries. The huge crushing machines would have destroyed this effective and efficient tradition of recycling materials, thrown a large community of people out of work, and produced solid cubes that were good for . . . what? All you can do is pile them up and make garbage pyramids in the desert next to the originals! And let us not forget that sometimes up to 90 per cent of 'foreign aid' is spent in the donor countries for making the recommended machines and materials. These funds actually provide employment for people in the donor country itself. How 'foreign' exactly is foreign aid?

Nevertheless, total incompetence and real 'rip-offs' are comparatively rare in consulting, and are not the main problem. The major problem is whether a consultant can give the best advice with complete openness, freedom and frankness without jeopardizing consulting possibilities in the future. For it is a simple fact that outside consultants are sometimes brought in not to give genuine advice in a difficult area, but to legitimitize preconceptions and policies which have already been made. 'Giving the client what he wants' is an old adage in consulting work, but it can raise those sorts of ethical tensions that can tear you apart. Because the issue is so sensitive (as most ethical issues are), I am going to illustrate the problem with an experience of my own, although I know of lots of others first-hand – some told to me out of deep concern, some narrated with blatant cynicism.

We have to go to Ghana in 1959, two years after Independence, when the road system was the finest in Africa. In the old Public Works Department morale was high as professional Ghanaian engineers took over from their British colleagues with feelings of mutual respect – the sort that come from working together on worthwhile things. Road maintenance schedules were strictly adhered to: they had to be both in the tropical rainforests of the south, as well as in the northern areas of torrential seasonal rains. The ubiquitous mammywagons, the small and efficient trucks, moved easily along the small laterite feeder roads as they 'smelt out' a load of cocoa here, a shipment of yams there. Exports were booming, and food was plentiful and cheap in the cities and towns.

Seven years later everything was a shambles. Road maintenance had been cut in half, and the funds diverted to other national tasks such as building triumphal arches and football stadia, constructing three presidential palaces (now hotels), and augmenting ministerial bank accounts in Switzerland. The first president, Kwame Nkrumah, had just decided it would be prudent to prolong his visit to China, and after the dancing and singing to celebrate the release of hundreds of political prisoners, the new government decided to take stock and rebuild. It was at that point that I was asked to go out for a month to assess the damage and recommend action for repair.

At the Accra airport I was whisked past all the formalities (for which I was grateful after a long airflight – no 747s in those days), and the next morning I had the first of several briefing sessions. The briefings, and the events of the next few days, were curious: *what*, exactly, was I going to recommend in my report? Since, in those first days after my arrival, I had no idea what the actual situation was in any detail, I could only say I did not know. After all, that was why I had come back to Ghana for a month – to find out, to learn, to think, and, as a last step, when all the facts were in, to recommend.

Ah yes, but wasn't it clear that many of Ghana's current transportation problems arose because there was not a main, modern (read American, four-lane, interstate) highway between Accra the capital on the coast, and Kumasi, the main inland centre? Frankly, the idea had not occurred to me, and in view of the virtual dissolution of the laterite feeder roads in the south, and the destruction of major laterite trunk roads in the north, it hardly seemed a high priority.

Oh, but obviously I had not understood: many of the problems would sort themselves out once a major highway was built. Would it not be a good idea to recommend such a joint enterprise? One could almost see the ribbons being cut, the flashbulbs popping, the symbol of black and white hands clasped in mutual endeavour.

I began to grow terribly uneasy. After all, the major alignment of parallel road and railway links between Accra and Kumasi had developed in a great arc over nearly sixty years. Major market towns had grown up on this historical alignment, and these would be severely affected by a 'straight line' route, much of which would pass through areas of sparse population. In brief, we were not talking about the long-range development of a Brazil, but short-term emergency action to get a country back on its feet.

Obviously nothing was going to be learnt or accomplished in

The major towns and lines of transportation in Ghana nine years after Independence in 1966. Lake Volta was still filling up, cutting major roads to the north, and many small feeder roads had become impassable.

Accra, and the Ghanaians kindly provided me with a car to see for myself. And see for myself I did: from a new road that had run out of funds in the far southwest; to quagmires of feeder roads in Ashanti; to deeply rutted trunk routes through Wenchi to the north, past Tamale and Bawku, and south again to the yam fields around Yenchi. It was the same pathetic story everywhere. You *cannot* maintain an open flow of commerce under tropical conditions – flows of food, materials, people, and the thousand and one things that move in a modern economy – *unless* you are prepared to allocate sufficient funds to maintain the roads. They have to be constantly graded; V-shaped brushes must pull the hardened laterite at the edges to the centre; culverts must be kept in good repair; ditches must be kept clear, or water quickly undermines the best-constructed earth courses. Unglamorous, unexciting, everyday, and ulti-mately *vital*, maintenance. No flashbulbs, no symbolic hand-clasps; just patient, slogging maintenance. It was not cheap, but it was fundamental. Maintain the roads, and your other development plans stand a chance. Fail in this most crucial of tasks, and the rest is jeopardized even before it starts. No one knew this better than the devoted Ghanaian engineers themselves, men who had watched their pride and joy fall apart in front of their eyes. Along the rutted roads, graders and tractors were rusting where they had stopped, because no spare parts were available, and no equipment was serviceable enough to pull them back to the workshops.

These were purely technical and engineering matters, but it was easy to see the human consequences. It was as though all the capillaries and small arteries of a human body had been blocked off – no feeder roads, no feeding. No bridge or ferry repair, no connections, no flow. Whole regions of healthy national tissue were turning gangrenous. Yam prices had risen fourfold at the same time that foodstuffs were rotting on the farms. Almost the entire crop around Yendi stayed where it was, at a time when this important seasonal crop was desper-ately needed in all the major towns, and in some areas bags of corn went up more than six times their former price. In a protein-deficient land, fish were rotting on the shores of Lake Volta. Fathers could no longer feed their families in the cities, and the children, and often the mothers, were sent 'home' to the rural farms. The men stayed on to hold their jobs and homes, only to live a lonely, alienated life with all its attendant psychological problems.

Administration was also being torn apart: agricultural exten-sion services contracted with the disappearance of small rural roads; community development officers could no longer infuse the same enthusiasms; and on-the-spot knowledge disappeared as field trips and visitations were curtailed from Accra and other regional headquarters.

After two weeks of travel, it was quite obvious that two things were needed: firstly, two-and-a-half million dollars worth of spare parts, much of which would be spent in Peoria, Illinois, to get the equipment back on the roads so the job of reconstruction could begin. The list had already been compiled by the Ghanaian engineers. All that was needed was the foreign exchange, the *real* help at that moment that could let the Ghanaians help themselves.

Secondly, plans had to go forward immediately to start transportation on the rapidly filling Lake Volta. The gigantic body of water had cut the former paved Kumasi-Tamale route, and had severed completely the yam region of the northeast from the markets of the south. After talking far into the night with sailors and marine engineers in Takoradi, the answer was simple: small boats of plentiful and fine tropical woods, expertly crafted in local shipyards, copper-sheathed against worm, with small, reliable Lister engines, diesels of proven worth that could keep going for years with minimal maintenance. Not, *decidedly not*, the system recommended by American consultants: a system demanding enormous quantities of foreign exchange, described in a two-volume, maroon-bound, gold-titled report with obsequious reference to the former President, containing plates of Mississippi River barges and other accoutrements of navigational overkill. What was needed was a 'mammywagon of the lake' to smell out the loads along a still-changing shoreline, a boat capable of pulling into a roughly constructed and temporary wharf or landing stage to pick up the same potpourri of people, goods and foodstuffs that the traditional land-based version collected.

Such views were not appreciated in Accra. Spare parts and lake boats were all very well, but what about the highway between Accra and Kumasi? This was the thing that really was the priority item – no?

No, I explained, this was hardly a priority now. The Ghanaians already had the basic equipment: they could do the job themselves with a bit of a leg-up. Spare parts were not very glamorous, but if we really wanted to do something that would be truly effective and appreciated, spare parts it had to be.

Yes, but wouldn't I, nevertheless, recommend the Accra-Kumasi highway after these more immediate things had been taken care of? I was non-committal, noting that very strong human geographic alignments would be devastatingly affected by such an enormous engineering project. Before anything like this was attempted, a full impact study, emphasizing the social as well as the economic dislocations, was essential. But it was clear that no one gave a damn about 'human geographic alignments'.

The real trouble came when my report was submitted to Washington. It was, frankly, unacceptable. Would I come to Washington to discuss it? Payment would be withheld until I did. A young bureaucrat, with a few weeks in West Africa under his belt, took me to task. What about the Accra-Kumasi highway? What about recommendations for a 'team' to study the question? I pointed out, once again, that the Ghanaians really needed spare parts, that they had the list virtually in hand, that no 'team' was necessary to tell them 'how we did it back home in Utah'. They also had a broad lake highway to the food-producing areas of the north, if only they could get boats moving.

But was I not going to recommend a new Accra-Kumasi axis? No, I was not – it needed much more study. Ah ha (pounce!), why then had I not recommended a 'study team'?

The problem was an exercise in non-sequiturial logic. How do you explain to someone trained in economics that if you do not recommend his goddam highway in the first place, that you don't need 'the boys from back home' to study it?

But to my everlasting discredit, my courage failed. It was the first bit of consulting I had ever done; perhaps they really could withhold payment for an 'unsatisfactory' report. So I put in an appendix in the approved style – a model was provided. Such-and-such a position for so many days; an economist, naturally, to do this-and-that for so many days more. They could study the thing until doomsday as far as I was concerned. It was obvious that the highway was going to be shoved down the throats of the Ghanaians one way or another. Perhaps a carefully chosen 'team of experts' would recommend the right thing. To my credit, I did not.

It was ten years later, ten years of shifts in personnel, attitudes, regions, countries and concerns, before I was asked to consult again. The circumstances were very different the second time: many things had been learnt, perhaps the most important one being that in the long run no one really gains by hiring people to legitimize preconceptions. But the difficult question remains for anyone who takes on consulting: do you do your best as you see it, give of your best skills, and write the clearest, most sharply reasoned report you can, even when you know that it may go against strong preconceptions, and so cut you off from any consulting again? Or do you assume a . . . well, shall we say a more mature and realistic stance, an attitude of compromise, giving the client what he wants, slipping in a bit of a recommendation here and there, just a suggestion perhaps, so that you will be asked again, and so have some leverage, some point of entry, to affect the process in the future?

Most people find this a very easy question to answer – either one way or another. Well, there is the double-edged sword again. Do you speak the truth as you see it, running the considerable risk of being cut out in the future? This means you cannot affect the course of developments even if you care deeply. And if your livelihood should depend upon your consulting activities, what do such pressures imply? Or do you try to figure out what 'they' want – and do your best to give it to them? The tension is always there, but it seems to me that people in the academic world have a special responsibility to profess, to the best of their ability, the truth as they see it. After all, they call themselves *professors*.

Postscript: Sometimes good things happen. It was years later in Sweden that a distinguished consultant to the World Bank and Ghana had come across my report. It was, he said, the best thing he had ever seen on the topic, and the Lake Volta transportation scheme had been adopted practically as I had recommended it. It was nice to know.

Part V
The geo-graphic revolution

The explosion in cartography

I have hedged my bets, and sat on fences, with that (r) in the geographic (r)evolution, but if there is one area where I am prepared to toss all caution to the wind, it is cartography. Here is an area that has literally exploded outwards, and has helped us all to understand the illuminating power of graphic presentation. It is a thoroughly geographic explosion, because from the earliest days maps have been associated with the geographic endeavour. To see this, let us do a little experiment together. We see someone we know, and go up to them without any warning, saying to them, 'What do *you* immediately think of when you hear the words *geography* and *geographer*?' They will probably be a bit startled, but prepared to humour us. Geography? Geographers? Well . . . I don't know. . . . Maps, I suppose. . . Try it, it works almost every time. Maps are indissolubly linked in most people's minds with geography, and you only have to riffle through the pages of this book to know that it has been written by a geographer, rather than an entymologist, a geneticist, a psychiatrist or an historian.

Not that people in these fields, and many others, hesitate to use maps when they think they need to. I have seen highly imaginative maps showing the intensity of wing flutters of a tethered moth as bat squeaks were generated at various locations on an imaginary globe surrounding it. One way to 'see' the flutters of agitation was to project them on to the globe, and then 'unfold it' to the familiar Mercator projection. In genetics today we constantly hear reports of 'mapping' a chromosome, and one distinguished scientist has described gene probing as the same as 'marking villages on a road map'. Even psychotherapy can use maps, as we shall see later. As for historians, the good ones are really brothers and sisters in arms. Historical atlases are part of the historians' trade too, and some even map such things as the goitres and hernias recorded by the recruiting teams of Napoleon's army. A fascinating story they tell too; of the heavy lifting by people in the port towns

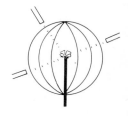

before machines, and the difficulties of supplying fish to the iodine-deficient villages of the Alps. Maps are by no means the exclusive property of the geographer, but few would deny that they have a very special geographic 'feel' to them. People expect geographers to be 'good at maps', and I personally think their expectations are justified. I also think that geographers have a responsibility to help others use maps far more effectively and with much greater imagination, for it is a constant and pleasant surprise to a geographer to see the way others react to cartographic representations that geographers too easily take for granted. Familiarity tends to breed contempt in all of us.

Let me give you a quite personal illustration. One morning I found myself talking to a committee of the Arab League about some research designed to help them understand the different sorts of television programmes that were flowing through that part of the world. We know very little about these artifacts of modern culture, even though these extraordinary combinations of sight and sound are some of the most potent forms of human communication ever devised. It was a hot day in Cairo, there had been lots of polite and rather general discussion before, it was getting near lunchtime . . . words alone would not do the trick. So I projected a map of the world on to the screen, a rather strange map based on priority telegrams between countries, rather than television flows (for which we had no data). It was a strange world, showing the relative locations of countries as points in 'press telegram space', and you could hear the rustle and the stirring as people half asleep sat up and lent forward to look. There was the First World (Europe, North America, Japan) at the centre of this particular 'information space', and there, in a remote peripheral ring, was the Third World. 'Suppose,' I said, 'suppose you could make a map like this for television flows, so countries connected by large flows would be close together, while those seldom using each other's programmes would be far apart? Would this be of interest to you? To see yourselves in television space? To see who was close to you, who was influencing you, and perhaps how you had moved over the years?' I really did not have to say much more, for the *map*, and the idea lying behind it that helped them think about themselves in a space they had never even considered before, did it all for me. Immediately they began to look for their own countries, to see what their part of the information space looked like. This is a common experience, for people often look for their own locations first, and then use them as personal anchor points

David Ley
University of British
Columbia
1947–

190

from which to move their thinking and imaginations outwards to other regions of the map.

A personal anecdote, yes; but I think that sudden rustling and stirring of interest on a hot Egyptian morning was indicative of something much deeper. We are the beings who see, and from our seeing we gain a powerful sense of certitude. '*Show* me,' says the sceptic, implying that once seen his scepticism will melt away. 'Oh, I *see!*' we say, when suddenly the light dawns and a difficulty or puzzle is resolved. And our language is impregnated with visual, map-like terms: we sketch out a *plan* and *map* out the next move; we speak of *regions* and *fields* of inquiry, suggestive of bounded spaces; we feel *close* to someone, or say they seemed very *distant* this morning. Perhaps one picture or map really is worth a thousand words?

So maps are pictures, but very special pictures because they often let us do an 'Oh, I see!' which was impossible before. This is because in a rather deep sense, maps are *models* – those abstractions, simplifications and compressions of the things we have chosen from 'out there'. Sometimes they are literally physical models like globes, or those plaster or plastic three-dimensional relief maps that so many people find fascinating. But whatever their form, they are all *filters*, because in constructing any map a geographer selects only a few things from the myriad of complexity 'out there', and filters all the extraneous details away. Sometimes it takes a great deal of imagination to select the filter that allows only certain things to get through. Have you ever seen, for example, a map of graffiti, those scatological slogans that range from the scratches on the walls of Pompeii, to the spray can pictures on subway trains in modern New York? David Ley and a colleague mapped the density of these in Philadelphia (Figure 17.1), and so disclosed the territories of the gangs who marked the boundaries of their turfs with slogans of defiance and insult, rather like wolves urinating to mark their territories – although wolves tend to be better behaved and seldom attack their own kind. Or what about a filter that lets us see machine space in a city, the actual area given over to machines instead of people? We always thought that the traffic was getting worse, but when we actually see the way acre after acre is falling to the machine, we wonder if human beings have any rights at all. Perhaps you thought I was exaggerating in Chapter 4? Do you now? Now that you can see the machine takeover with your own eyes?

Unfortunately, even simple but provocative maps like these may take many hours to compile. Not simply to collect and compile the information, but to locate it carefully, and only

191

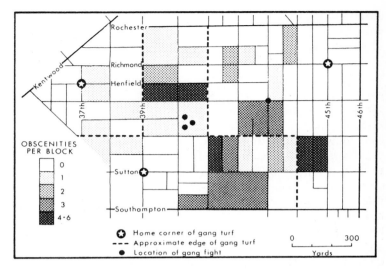

Figure 17.1: A portion of Philadelphia indicating the density of graffiti – the slogans used by gangs to mark their territories and assert themselves against their neighbours.

then carry out the design and final drawing of the map. Today, however, the computer has changed all that, and now automatic plotters do in a few minutes what may have taken hours or even days before. At the same time, letting the computer do much of the tedious work has brought to our attention a curious omission in the way we traditionally collect information. I say a 'curious omission', because I think it is indicative of society's deep failure to consider critical geographic aspects of problems. When we take a census, we are always very careful to locate it in time. A national census does not take place in a casual fashion, so that we gather information about people in this area in January, and that area in March. On the contrary, everything is geared up with an enormous amount of previous planning, and at midnight on 31 March off we go. So why when we locate information so accurately in *time*, do we not see the necessity of locating it with equal precision in *space*? It would be so easy to do: simply set up a national grid, and locate all census data to the nearest 10 metres. Then we could make maps of anything we wanted, at any scale we needed, at the touch of a computer button. We would also get around the bane of every geographer's and planner's life – the fact that the political units we now use to collect data in are not only of different sizes, but change so often that it is practically impossible to make direct and detailed comparisons between one census and another. If you wanted to design systems that were as geographically useless as possible, you could not do much better than the ones we have now.

Is anyone doing any better? Yes, and as usual we have to

192

go to Sweden to see some sensible geographic thinking at work. Even here it took nearly twenty years for the idea of geocoding to penetrate the traditional bureaucratic mind, from the time it was suggested by Torsten Hägerstrand in 1955, to the time it was really operational in the 1970s. By simply attaching x and y coordinates to each piece of data – a *datum* – information can be aggregated and mapped at any scale that seems appropriate. You want to build a new kindergarten? Fine: Swedish topographic maps can be overprinted with one kilometre grid cells by computer, recording the numbers of children who are to be served. Too fine a grid? Too coarse? Very well, tell the computer what size of grid system you need, and it will produce it for you. After all, why not? It has information down to the nearest 10 metres. In fact more and more scientific mapping is being carried out in this way today. Maps in botanical atlases are produced by computer, as well as maps showing the dry depositions of sulphur that pinpoint where dilute acids eat away building materials and destroy millions of conifers.

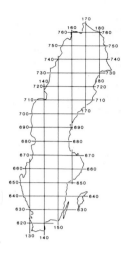

But the computer does more than simply scan its files and compile data by coordinates, important though such quick and flexible abilities are. With our computer graphics today, we can lift mountains, putting them on our computer screens, and twirling them around for our inspection until we get just the view we need. Sometimes we take the masses of numerical data we obtain from stereoscopic pairs of photographs, and let the computer draw the three-dimensional map we want (Figure 17.2). In this way, Mount St Helens stands revealed, both before and after the massive volcanic explosion that ripped a quarter of the cone away. Or, to anticipate the next chapter a little, we may take our data from satellite sensors, and let the computer reveal the crumpled corrugations of the ridges and valleys of Pennsylvania (Figure 17.3). It is almost impossible to imagine such maps being drawn by hand only twenty years ago.

Hippuris Vulgaris

In a sense, however, all the examples we have looked at so far are familiar and still fairly conventional, although do not underestimate the power of simple maps to reveal things we did not see before, or to raise questions we did not think of asking. As we shall see (Chapter 19), such maps become very important when geographers start collaborating with people in medicine. Nevertheless, we are still on fairly familiar and conventional ground. The really big change in mapping, the essence of the cartographic explosion, lies not just in doing things faster, but in doing them differently, in ways we never thought of doing only twenty years ago. It is the explosion

193

Figure 17.2: Mount St.
Helens in Washington
State, before and after
the massive explosion
that tore out a whole
side. Computer drawn
relief maps from
stereographic pairs of
photographs.

Figure 17.3: A
computer drawn map
of a part of central
Pennsylvania
employing digital
terrain modelling.
Notice the strong three-
dimensional effect of
the famous 'Ridge and
Valley' section
enhanced by the careful
selection of the
viewing angle.

in cartographic *research* that has changed the face of mapping, and if Chapter 12 on human contacts belongs to Torsten Hägerstrand, then this chapter really belongs to Waldo Tobler. Almost single-handedly, from his days as a student with William Garrison to the present, he has made nearly all the running, and has inspired a new generation of young cartographers to follow the path of creative and imaginative map-making.

Some of his approaches are deceptively simple, the sort that make you say, 'Why on earth didn't I think of that?' Take something as apparently obvious as comparing two maps, say a very old one like the map of the British Isles hanging in Hereford cathedral, and the map we have today meticulously surveyed by modern instruments. How do we compare them, and can we learn anything from such a comparison about the way cartographers employed projections in the fifteenth century? Perhaps we should have a stage aside here on projections, those mathematical tricks that geographers use to turn a

Waldo Tobler
*University of California,
Santa Barbara*
1930–

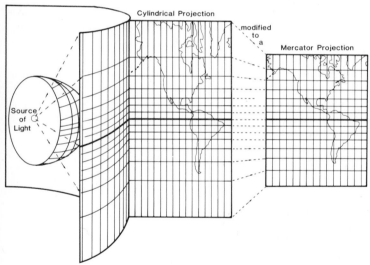

Figure 17.4: A source of light inside a translucent globe 'projecting' the image of the continents onto a cylinder that touches the globe at the equator. When the image is fixed, and the cylinder slit down one side and unrolled, we have the familiar Mercator Projection.

round globe into a flat map. They are well named, because we can literally think of a light shining through our translucent globe, *projecting* the shapes of the continents on to another surface, perhaps a cylinder (Figure 17.4), or a cone. If we fix the image photographically, we can cut the cylinder or cone and unroll it on to a flat surface. In the case of the cylinder, we would have the familiar Mercator projection, and you can see now why Greenland, far from the Equator where the cylinder touches the globe, has been stretched to make it appear as big as South America. If we want less distortion, say

195

for an area the size of the United States, we would probably project our globe on to a cone, letting the cone cut our globe at two parallels of latitude to keep it as close to the surface as possible, and so minimize the distortion when the light projects the shapes on to the cone. Of course, we do not really do it this way, but use mathematical transformations, all carried out today by computers and automatic plotters.

But back to Tobler's problem of comparing maps, and let us think about it at a purely practical and intuitive level. We have two maps, an old and a new one, and we want to see how well they 'fit'. What is the first thing we would do? Probably put them on top of one another, and then hold them up to the window, or lay them on a light table (a table with a glass top and a source of light underneath). Then we could move one map around with respect to the other to see how 'close' we could get them. In juggling the maps, we could actually do three things; we could shift one up or down, or right and left, something a mathematician would call a *translation*. Or we could turn one map with respect to the other, what a mathematician would call a *rotation*. Finally, if we had a camera-like affair called an enlarger-reducer, we could *scale* one map, making it bigger or smaller to fit the other. I think you can see that if one map were simply a 'reduced-upsidedown-shifted left' version of the other, we could 'undo' it and make it fit perfectly by enlarging it to the correct size, rotating it upright, and shifting it right. Then one map would lie exactly on the other, and the fit would be perfect.

In the case of the old maps of the United Kingdom, we cannot make it fit our modern one no matter how hard we try, although we can do all these shifting and enlarging operations mathematically until we get the closest fit possible. Unfortunately, the fit is still not perfect, and here is where two important questions arise. Firstly, can we disclose where the two maps still do not fit, the places we could stretch and squeeze if one map were drawn on a rubber sheet so that we could deform it locally? Secondly, could we do this by a simple stretching or squeezing, something a mathematician would call a *continuous transformation*, or would we have to *tear* or cut up one map to make it fit the other? Today, the computer does all this for us, trying to get the best fit it can, and then it plots the amount of local distortion as a series of arrows whose lengths tell us the amount of stretching and squeezing we still have to do on our rubber sheet. Moreover, if the arrows (we actually call them vectors) cross, it means that there is a really serious dislocation that can only be overcome by cutting or

196

tearing the map. The vectors really tell us the places where the two maps do not fit very well, and in the case of old maps they give us a clue as to the types of projections the old map-makers might have used.

With these cartographic tricks, geographers get up to all sorts of things. When women play field hockey, they pass the ball to and fro, trying to maintain close-knit connections between themselves, working their way up the field, until they get into their opponents' goal. In the meantime their opponents are trying to break up those passing connections, and so form tightly knit structures of their own. We have all read such things as 'Team X fell apart in the 4th quarter', and we can almost visualize poor old X broken up and scattered all over . . . well, what? Hockey space, of course! If we collect data about passes in our now familiar interaction matrix, whose rows and columns are the team members, we can make a map of our team in hockey space (a space *defined* by the passing relations between our players), and ask how does it compare to the formal and prescribed positions on the field (Figure 17.5). By translating, rotating and scaling one pattern, can we make it fit the other? In the case of one well-known team (Figure 17.6), the vectors of two players crossed, indicating that one player was far from the conventional prescribed pattern, a dislocation that could only be 'put right' by tearing the rubber sheet map.

Or what about those ideal central place patterns, with their neat little hexagons? Do they actually fit a real pattern of human settlement (Chapter 9)? In the Teotihuacan Valley of Central Mexico, one archaeologist asked whether an old pattern of late Aztec settlement could be continuously transformed into the lattice-like structure predicted by the theory. The answer was yes (Figure 17.7), and the rubber sheet vectors

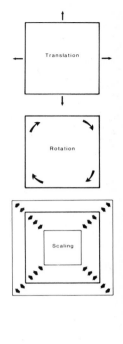

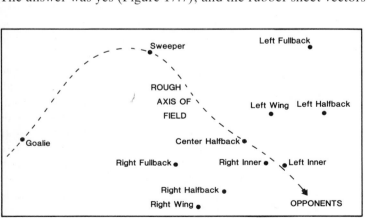

Figure 17.5: The actual positions of a women's hockey team derived from the way they pass the ball to each other.

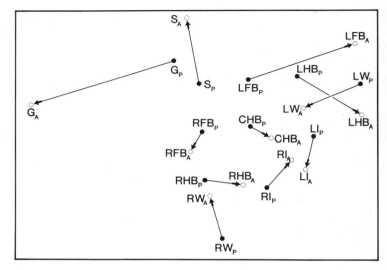

Figure 17.6: When one map is translated, rotated and scaled to fit the other, there is still local stretching and squeezing to do (vectors), and the fact that two of these cross indicates that ordinary field space has to be torn and ruptured to make it fit hockey space perfectly.

disclosed how the purely theoretical settlement space had been distorted by political, economic and physical factors. The small places had been drawn to the fertile and well-watered valley, and pushed to the boundary of the area by the massive intrusion of El Gordo, a large volcano. In contrast, the large places (Figure 17.8) had either been pushed to the eastern boundary to form strong administrative centres in a defensive marchland area, or pulled close to the lake shore – a very strong connection to the rest of the Aztec 'system' which had boats but no wheeled vehicles.

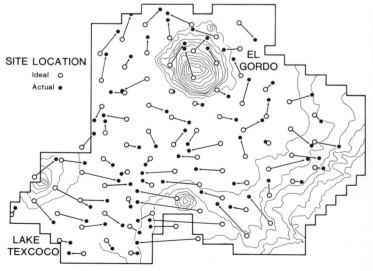

Figure 17.7: The small central places of the late Aztec settlement pattern which can be continuously transformed into the hexagon-like lattive predicted by theory. Notice how they have been drawn to the well-watered and fertile valley, and pushed outwards by the intrusion of the volcano El Gordo.

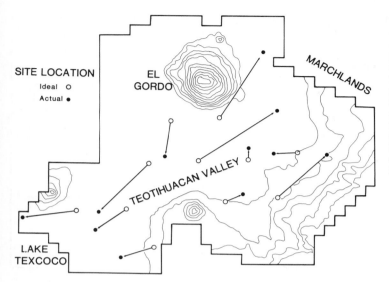

SITE LOCATION
Ideal ○
Actual ●

EL GORDO

MARCHLANDS

TEOTIHUACAN VALLEY

LAKE TEXCOCO

Figure 17.8: The large central places of the late Aztec settlement pattern continuously deformed from the theoretical hexagonal pattern by the political need to establish strong administrative centres to the north-east, and the economic need to be connected to the rest of the Aztec system by lake boats.

Have you ever thought what medicine, zoology, art history and physical anthropology have in common? They are all tied together by geographers' maps, particularly maps of skulls. When a brain surgeon has to operate, she needs the best and most accurate three-dimensional maps that modern CAT scanners can provide, and these have highly precise coordinate systems to locate the areas that are diseased. Zoologists have been using such maps for years, and comparing skulls (Figure 17.9) is really no different from comparing maps. Tobler's

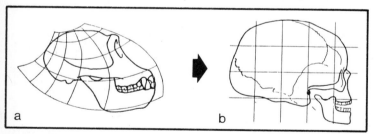

a   b

Figure 17.9: The skull of a chimpanzee (a) and a human being (b) shown in map form with similar coordinate systems suggesting how one might be transformed into the other.

rubber sheet vectors show us exactly how much of a 'stretch' it is between ourselves and our chimpanzee ancestors. Not much, it turns out. As for artists, we know that many have been fascinated by heads and faces, and whether it is an Albrecht Dürer (Figure 17.10), or someone working with a modern police identikit on a television screen, both are concerned with the way one human face can be transformed into another, a process that always involves stretching and squeezing and transforming a map.

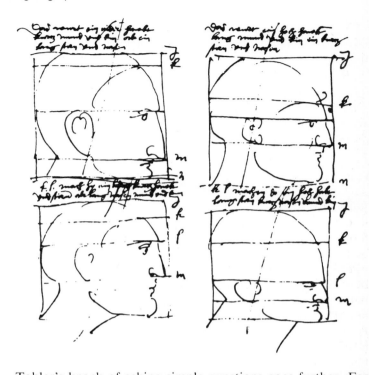

Figure 17.10: The facial
transformations of
Albrecht Dürer, who
worked considerably
before the days of the
computer (1478–1521).

Tobler's knack of asking simple questions goes further. For years, geographers were trying to figure out how to break up continuous values of data in order to plot them on a map. For example, if we have wheat yields by counties in the American midwest, how many categories and different gradations of shading shall we use? Shall we have three intervals or five? Shall we break up the values of bushels per acre in continuous steps, or increasing steps, or wherever we see 'natural breaks' in the distribution? Whatever intervals and categories we use, we will create a different cartographic 'text', with a different visual appearance, and so our interpretations may well change – as people making maps for propaganda purposes well know. You have no idea how geographers used to jump up and down and get all hot under the collar about this question! 'But wait a minute,' said Tobler, 'we have computers now with automatic plotters that can produce shaded patterns of any reflectance between pure white and pure black that you need. Why bother to break up lots of data into a few categories at all? Why not simply let the automatic plotter produce the exact reflectance value to match your bushels per acre, or whatever other things you are mapping?' 'Oh,' said the conventional cartographers, 'we didn't think of that.'

200

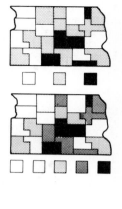

And that, of course, is often the trouble in any field. A new technical breakthrough opens up all sorts of exciting possibilities, but our thinking is too often trapped back there in the neolithic age, when maps were scratched on rocks. It is this quality of sensing the new possibilities that marks a creative and imaginative research worker. As my favourite poet put it:

And they asked me how I did it, and I gave 'em the Scripture text, 'You keep your light so shining a little in front o' the next!'

We have met Tobler's 'winds of influence' maps before (Chapter 13), but these can be made in any space and at any scale. Suppose we record the number of times geographic journals footnote each other, recording such acknowledgments in our usual interaction matrix. We may expect that journals footnoting each other often will be rather similar, so they will be represented as points close together in footnote space – and *vice versa*. It is unlikely that such intellectual interaction will be symmetrical; Journal A may footnote Journal B 43 times, but may only be footnoted 23 times in return. This means a net 'flow' of 20 intellectual acknowledgments goes from B to A, so perhaps B is a pretty high pressure journal containing lots of ideas. When we construct the map for 1975 for the major geography journals in English (journals in other languages would form their own clusters very far away, which says something about the language barrier today – see Chapter 27), we see a zone of high pressure in the centre of the space dominated by the journal *Economic Geography* (Figure 17.11).

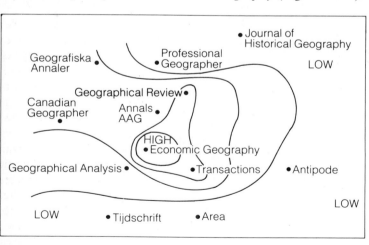

Figure 17.11: Some geography journals in footnote space in 1975. A zone of high pressure lies over *Economic Geography*, and from this central area intellectual acknowledgments flow to journals in low pressure areas on the periphery.

This highest value declines to more moderate levels in the *Canadian Geographer* and *Geografiska Annaler* (Swedish), and from this ridge of pressure the winds flow down to the *Journal of Historical Geography* (still new), *Area* and *Professional Geographer* (smaller journals of commentary), and *Antipode* (a journal of radical geography). By 1980, both the configuration of the journals and the pressure surface have changed (Figure 17.12). Now a ridge of high pressure extends from

Figure 17.12: Geography journals in footnote space in 1980: not only have journal locations changed (notice how *Transections* and the *Geographical Review*, for example, have moved to the periphery), but the former zone of high pressure at the centre has shifted to a ridge that includes *Geographical Analysis* and *Antipode*.

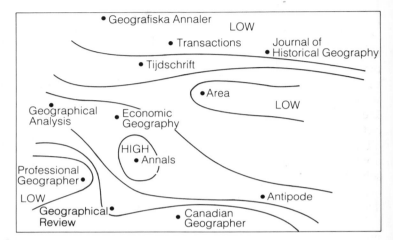

Antipode (which has become a net source of intellectual acknowledgments), through the peak over the *Annals*, to the theoretically oriented *Geographical Analysis*, and again the winds of ideas blow from the net generators of ideas down to the net receivers. These are still *Area* and *Professional Geographer* (which play an important commentary and review function), and *Geografiska Annaler*, which was formerly a high-pressure source. Perhaps we can imagine geography as a slowly moving ballet, with journals as dancers constantly pushing up or depressing a pressure surface over them that causes the winds of ideas to flow this way and that.

A large part of the cartographic revolution is a result of geographers wrenching their thinking out of conventional geographic space, and letting their imaginations, and those of people in other fields, roam in other spaces, such as those defined by academic subjects (Figure 5.1), hockey players, and professional journals. For space is not a wastepaper basket that sits there waiting for us to fill it with things, but something we define to suit our needs. A psychotherapist, for example, might record the progress of a patient who was having difficulty coping with the world, yet what does 'progress' suggest except going from 'here' to 'there', moving along a path in psychotherapy space (Figure 17.13). One young woman felt so inad-

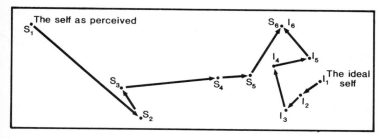

Figure 17.13: The six-week journey of a young woman in psychotherapy space. Her 'self as perceived' ($S_1 - S_6$) moves a long way, but her 'ideal person' ($I_1 - I_6$) also moves closer to her own self-perception.

equate that her image of her self (S_1) and her image of an ideal person (I_1) were far apart during the first week of a course of therapy. But over time she regained confidence, and after six weeks her perception of her self (S_6) moved much closer to her Ideal (I_6). Notice that the location of her Ideal also changed, perhaps because she became a bit more 'realistic', for we can all hold up such impossible standards for ourselves that we are bound to fail. After six weeks of help, her perceived and ideal selves were much closer together, and she was 'discharged', a much happier person able to cope once again with the ups and downs of her life.

Even in literature, the geographer's map brings out some provocative patterns and questions. If we simply record who speaks to whom in *Romeo and Juliet*, we can make a map of the play in what we might call Shakespeare space (Figure 17.4).

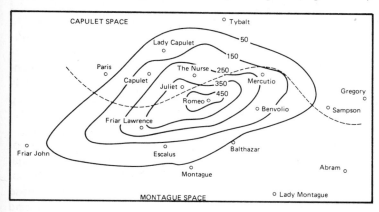

Figure 17.14: The map of *Romeo and Juliet* in Shakespeare space. Romeo, with the most lines, is in the centre, on top of a sort of verbal hill, with Juliet and the Nurse not far away. Lesser players are downslope towards the periphery. The area divides neatly into Capulet and Montague regions.

Romeo is in the centre (he never could keep his mouth shut), but Juliet is not far away, close to the Nurse, who so often acts as an intermediary between the two lovers. Other players are closer to the periphery, or 'lower down the slopes' if we draw in contours indicating the total number of lines they speak, and notice how the space divides into two distinct Capulet and Montague 'regions'. A number of Shakespeare's plays have been mapped (in *Hamlet* all the dead at the end of the play end up in one region, all the living in another), and they display great, but always highly interpretable, variety. In contrast, the plays of the great French playwright Racine display almost exactly the same pattern (four principal characters in the centre, four lesser ones on the outside), and it seems legitimate to ask if he was writing to some sort of formula. Saying this does not in the least denigrate the genius of Racine, but perhaps these maps do come close to disclosing something that we rather vaguely call the 'structure' of the play?

These approaches to map-making in strange and new spaces apply just as much to things that really are located in geographic space, although we may not 'see' all the implications until we have transformed the old and familiar map in appropriate ways. A conventional map of Canada is all that is needed to locate its major towns, but suppose we transform Canada into newspaper information space (Figure 17.15). Right at the centre (and

Figure 17.15: Major cities of Canada in newspaper information space based upon how cities report each other. Toronto and Ottawa are so close together they have to be shown as a single point. Vancouver, far to the west in geographic space, has been pulled closer to the centre.

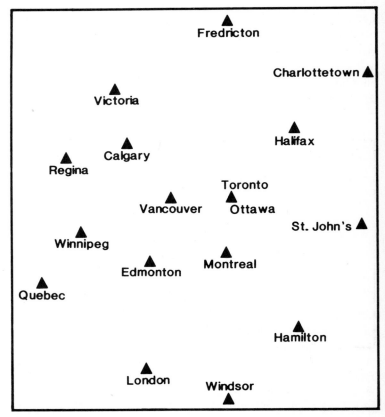

recalling what we know now about accessibility) are Toronto and Ottawa, with Montreal not too far away, although we might speculate that Montreal has moved farther out to the periphery and closer to Quebec these days. This raises, of course, the question of trajectories in these spaces, the paths traced out by the movements over time that perhaps only an animated map could capture properly. Other towns, quite close to Toronto in geographic space, are far from the centre in this information space (perhaps we have a cartographic definition here of 'boondocks'), and look what has happened to Vancouver! Far out on the West Coast, it has been subject to a wrenching, tearing transformation bringing it quite close to the centre of the country.

'But this isn't the *real* Canada', some might say. 'It's only a picture from the cartographer's imagination.' Really? Do you believe that? What *is* the reality of Canada today? The old conventional map that traps old conventional thinking? Or the shifting dynamic map of information space that is so humanly relevant? Which 'reality' helps you understand modern-day Canada better, which map illuminates the current Canadian condition, making you say, Hm, . . . I never thought about it *that* way before?

It is the same in Europe: What is the *real* map today? The familiar one hanging on the schoolroom wall, with Sweden looming huge as it is stretched by the projection? Or this map of Western Europe in postal space (Figure 17.16), showing the

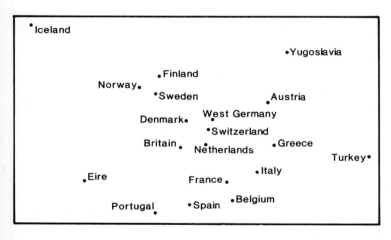

Figure 17.16: Europe in postal space. The 'core' seems to be West Germany, the Netherlands and Switzerland, with Britain and France not too far away. Is Denmark slowly making a journey from her old Nordic partners towards the Common Market countries?

countries located according to the letters they exchange? What is the *centre* of this Europe today? There they are: West Germany, The Netherlands, Switzerland, and, not too far away, the United Kingdom. Is this surprising when 80 per cent or more of all mail flows are business and banking? There, too, is France, not quite at the centre, but at the middle of a crescent of countries – Italy, Spain, Portugal, Belgium – all sharing the Latin base of their languages. And look at Denmark, moving ever closer to the centre, and farther and farther away from her traditional Scandinavian partners, in marked contrast to Norway, still solidly with Sweden and Finland. This is particularly intriguing when you think that Denmark voted to go into the Common Market, and Norway voted to stay out. As for the periphery, Turkey is on one side with Iceland on the other, and Eire is being pushed away by its lack of interaction with the rest of Europe, yet trying to get close to the United Kingdom with its traditional commercial ties. We can almost feel the tension in this part of the map, the contradictory push-pull forces at work. Once again, is this a figment of the cartographic imagination, or is this *a* reality we should try to understand? But before we can understand, we have to *see*.

Finally, these new ways of map construction, in all their computer complexity, may help us recreate landscapes of human commerce and interaction now long since disappeared. Babylonic cuneiform tablets, incised by the stylii of scribes thousands of years ago, mainly recorded commercial transactions – so many jars of oil, wine, incense, and perhaps even gold, frankincense and myrrh. If we collect hundreds of these

Figure 17.17: A map of ancient cities in Babylonic cuniform space, created from the commercial transactions recorded on clay tablets. The map predicts the locations of many unknown sites, and perhaps one day archaeologists will tell us if we are right or wrong.

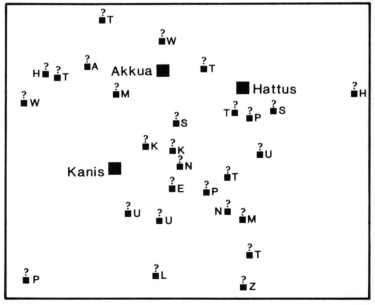

transactions in our interaction matrix, we can create maps of towns as they were 3,000 years ago (Figure 17.17). This was the Cappadoccian speculation of Tobler, and the strange thing is that the three towns we know from archaeological excavations are just about in the right position relative to each other. What about the others? We do not know, because they are still lost cities, but a geographer might well say to his archaeological colleague, 'If I were you, I'd have a poke around there – you never know what you might find.'

Computers and plotters, transformations and strange spaces – the cartographer's world has certainly exploded, forming an important part of the story of modern geography. But we have not forgotten the old spaces – both the inner space of the earth's surface, and how we view it now from the outer space made available to us by earth satellites. We are now sensing the earth, our little blue sphere hanging there in the darkness, and this outside-looking-in perspective also forms an important and invigorating part of the story. We call it *remote* sensing, but I wonder if it really is?

18 Not so remote sensing

When our latest caravels in outer space round Cape Saturn and spread their photoelectric sails for Neptune, when they send us signals of one *billionth* of a watt over hundreds of millions of kilometres, signals that we transform into pictures of volcanoes on Io, new moons of Uranus, and perhaps new planets beyond Pluto, then I think we can legitimately use the adjective *remote* to describe our sensing. In contrast, our earth satellites, moving mere hundreds of kilometres overhead, have a cosy, even local feel to them, mere dinghies probing the local bays and inlets around our home. That we should view them in this way today is testimony to the revolution in thinking we have all undergone in the last twenty-five years. It was only in 1957 that the little beeps of Sputnik announced the possibility of a new way of seeing and sensing our planet. Today we are beginning to worry about the traffic jams of rocket debris floating around in orbit.

How do we get images of our earth, and how do we extract information from them? Like many good ideas, the basic principles are very simple and straightforward. But like many good ideas their execution is technically difficult and horrendously expensive. Not just millions, but tens of billions of dollars, roubles, pounds, yen, francs and marks have underpinned our new found way of seeing, and these huge sums of money almost guarantee that the military justifications have been there from the beginning. And I mean from the beginning: observations were made from balloons by the French in 1794 to help them win the battle of Fleurus; aerial photography was developed extensively for military intelligence during the First and Second World Wars; and today constant and mutual surveillance by satellites constitutes a form of global voyeurism not without its prurient interest. As the old spiritual put it:

> I went to the rock to hide my face,
> And the Rock cried out 'No hiding place!
> There's no hiding place down here!'

So do draw the curtains, please.

From balloons to satellites, all forms of sensing have had one thing in common, whether human eyes, photographic films, or magnetic tapes were the initial receptors. All sensed radiation of one sort or another, is either reflected from the earth, or even transmitted – as a remarkable satellite photo of Europe taken at night shows (Figure 18.1). If you look carefully, you can

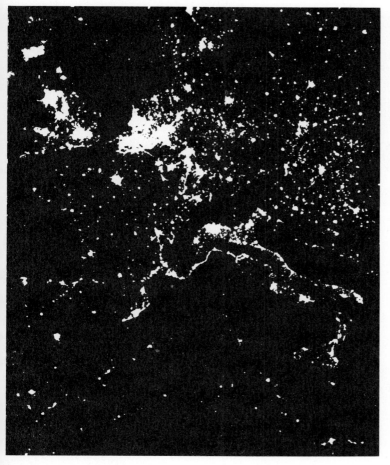

Figure 18.1: A satellite photo of Europe taken at night using the transmitted electric light. The lights of the coastal towns pick out the outlines of the continent, and the mountains and rural areas are great black voids. The 'core' of Europe in northwest Germany, Belgium and Holland is easy to pick out.

see the dark curving mass of the Alps blending into Italy's mountainous spine of the Apennines; the coastal outlines are also easy to pick out, but look at the blazes of those oil rigs in the North Sea! And there, once again, is the core of population potential we saw in Chapter 13, reaching up the Rhine Valley, and slopping over to London and the British industrial areas. Notice, in contrast, how empty France appears: there are a few splodges of light at Paris and down the Rhône Valley from Lyon to Marseilles, a few regularly spaced dots here and there

209

Gamma Rays
X Rays
Ultraviolet
Visible
Infra-red
Microwave
Radio Waves

Violet
Blue
Green
Yellow-Orange
Red
Infrared

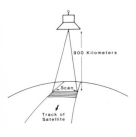

(recall central place theory), but much of the country is dark, the small lights of the villages scarcely penetrating the blackness of the night. France, for all her urban sophistication, is still a remarkable rural country – as her farmers and sensitive politicians know well.

Nevertheless, this extraordinarily provocative photo made by transmitted light is an exception to the rule: most images are made in daylight, from reflected solar radiation at a number of different wavelengths in the visible and near infra-red range. If we take the images made by the Landsat series of satellites, launched from 1972 to 1983, the reflected light is recorded on 4–6 different bands, usually blue, green, red and infra-red. In the satellite itself, about 900 kilometres above the earth's surface, a small mirror oscillates back and forth sweeping out a scan of about 185 kilometres. The radiation (visible and infra-red) is filtered into separate bands, either for immediate transmission to earth, or to be stored temporarily until more suitable conditions for transmission appear. Each scan actually records the average radiation from small rectangular areas 57 × 79 metres called pixels, and 2,340 × 3,240, or 7,581,600 of these go to make up one image covering 34,140 square kilometres. The sensing is going on most of the time in daylight, so you can see why we need large computers to store and handle the data. Since each pixel can record about 100 intensity values over four wavelengths, and each image contains 7.58 million pixels, we have to have the capability of handling enormous amounts of information. An image is actually transmitted to us as a long stream of numbers, and it is from these values that we have to try to reconstruct something we can actually see.

One way of doing this is to record the numerical values as varying intensities in each pixel, and then make a separate colour image of each band. The basic colours (usually blue-cyan, red-magenta and yellow), do not have to bear any resemblance to what we normally see, because streams of transmitted numbers do not bring any colours with them from outer space. We choose our colours to highlight the important features, and what we consider important varies with the particular task at hand. For example, if we were making a survey of wetlands – those coastal areas that we realize today are such a crucially important part of our ecosystem – we might choose to accentuate turbid water carrying sediments, swamplands, and various forms of vegetation. As Ronald Eyton has shown in South Carolina, these images, properly and thoughtfully reconstructed, can carry literally vital information for the biologist,

210

planner and politician. Unfortunately we cannot show his beautifully reconstructed images here in colour, but when they are overlaid to produce many colours and shades, they allow the trained observer to pick out a dozen categories of important land and water conditions.

Ronald Eyton
The Pennsylvania State University

Unfortunately, and no matter how we process the streams of numbers, the information is still quite crude, and no matter how much we enlarge such images we can never 'see' any details smaller than those resolved by the 57 × 79 metre pixel, although sharp edges and lines (roads and geological faults) do have a way of standing out sometimes. Even the newest Earthsat, which has a pixel resolution about four times as sharp as the old instruments, is still a fairly blunt instrument compared to those we are capable of constructing. Some say that our latest instruments can pick out the badges of rank of soldiers on manoeuvres, which is really quite a trick because soldiers do not wear visible badges of rank on manoeuvres these days. The resolution problem, as you might have guessed, really comes down to one of military secrecy, although I cannot help feeling that security classifications are taken to quite idiotic extremes. If U2 flights in the 1950s could pick up the white lines in parking lots, it does not need a vivid imagination to guess what we can do now thirty years later. It is ironical that the Russian and American intelligence services know the details of each other's capabilities far better than their own civilians do. Perhaps we have the purest essence of government bureaucracy here, when some agencies (the departments of military intelligence) require on security grounds the deliberate degrading of image quality, while others engaged in peaceful civilian use spend literally millions on virtually useless statistical exercises to extract information from images that have been deliberately blurred. Often the same agency is both the image-degrader *and* the disburser of image-enhancer funds. You wonder sometimes whether the word 'intelligence' can ever be legitimately employed in these areas. The unfortunate consequence of this schoolboy nonsense is that academic people in remote sensing are always working fifteen to twenty years behind the state of the art. It must be terribly discouraging.

The problem of image resolution is that the radiation measures from the 57 × 79 metre pixels are simply average figures, and from these values over 4–6 wavelengths we have to try to figure out what the 'ground truth' is. At first, the problem looks reasonably straightforward. Suppose we take three pixels whose 'ground truths' are vegetation, soil and water, and we plot their reflectance values over the violet to infra-red spec-

211

trum (Figure 18.2). Each type of 'land use' generates a different curve called its *spectral signature* – in a sense, its own quite distinctive thumbprint. If we take two or more bands (math-

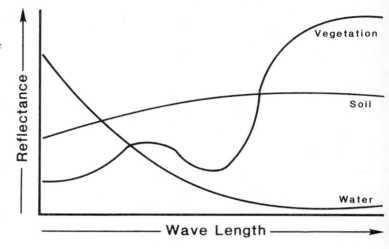

Figure 18.2: The reflectance values of different types of land use, measured over the violet to infra-red part of the spectrum, form distinctive curves of 'spectral signatures'. Those for bare soil, vegetation and water are quite distinctively different.

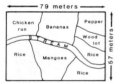

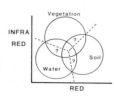

ematically we may take as many as we like), say red and infra-red, and plot the reflectance values, then our vegetation, soil and water pixels form distinctive clusters that correspond to the different classes of landuse (Figure 18.3). Everything seems quite straightforward in theory, and off we go.

But only in theory: unfortunately, in practice things are often not so simple. It is all very well if we have vast expanses of fishing grounds off the coast of Peru, or uniformly ploughed fields in Kazakhstan, USSR, because the average reflectance values of our pixels will give us clear spectral signatures characteristic of water and soil. But suppose the landscape is 'plotted and pieced – fold, fallow, and plough', a hodge-podge of different land uses or crops? Then our big crude pawprint of a pixel is going to slop over all sorts of different land uses, average all the reflectance values, and give us . . . a large electronic mess to sort out. And messes, by definition, are *not* information. The problem is that our neat little textbook example of clusters of vegetation, soil and water pixels (Figure 18.3) is often a big sloppy mess of overlapping nebula-like areas, and *somehow* we have to try to disentangle these and decide which is which.

Oh, no problem, says the statistician. You want to discriminate? Fine, we've got mathematical techniques called 'discriminant analysis'. And off they go with their computer programs, chopping up overlapping globules of pixels with

212

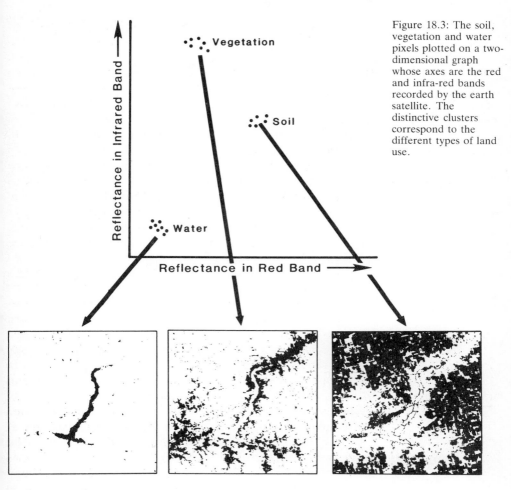

Figure 18.3: The soil, vegetation and water pixels plotted on a two-dimensional graph whose axes are the red and infra-red bands recorded by the earth satellite. The distinctive clusters correspond to the different types of land use.

straight lines, telling us that pixel number 1346 is *probably* soil (right in the middle of the lake?), and feeling very pleased with themselves. Well they might: *millions* of dollars have been spent by them on these simplistic approaches, and too often they give the lay person a feeling that the problem is being solved, when really it is being swept under the rug. In the meantime, ignorance, pretentiously clothed in inappropriate mathematics, is being substituted for real knowledge.

The *real* problem, of course, is that our civilian data have been deliberately degraded to prevent the passage of information, and no amount of statistical massaging can put back information that is not there. The only real solution is to try to obtain more information, and the best way is to stop fooling around with images twenty years behind the state of the art, and demand decent data in the first place. Another way is to scan an area two or three times (so adding to our information), and see what changes have come about. This is particularly useful for telling different crops apart, because many grains like winter wheat, rice, barley and so on, produce rather distinctive changes in their spectral signatures, particularly during certain phases of their growing season.

Those are the difficulties – at least some of them – but there are clearly great gains too. Whatever the technological gap between what we can do and what we are allowed to do, few can deny the exciting possibilities opened up by various forms of sensing. And to point to military uses is simply to point once again at the truism that all knowledge can be used for good or ill – and that your 'good' may well be my 'ill', and *vice versa*. For example, special camouflage detection film may disclose dying vegetation covering a gun position, but it can also pinpoint diseased and dying trees in an orchard. Tank engines may give off heat, and radiate their images in the infra-red (every American and Russian countdown and white-hot rocket ignition is also monitored by satellites), but infra-red sensing can also pinpoint buildings with excessive heat losses, and so help us save energy. Even the heat given off by illegal whisky stills in the hills of Kentucky and Tennessee can be spotted in the infra-red, although this, I feel, is not playing the game at all. It is rumoured that 'newer methods of distillation' avoid detection, and we can rejoice that the descendants of Scots and Irish whisky makers were around a long time before satellites.

In a sense, the possibilities opened up by remote sensing seem to be more limited by our imaginations rather than technical problems. This has been a constant theme of David Simonett, one of the geographic pioneers of remote sensing. Even before the first remote-sensing satellites were launched, he was pleading with geographers to think about what information they might want. I remember vividly a NASA conference in Houston, Texas, in the early 1960s in which we tried to respond to his pleas for more imaginative thinking, no easy task if you have never thought in this way before. I also recall the suggestion that chemical sensors on satellites might pick out the enriched soil from the droppings of camels in the Gobi Desert, so enabling historical geographers to reconstruct the old caravan routes. Since this is a book for all the family, I shall not record for posterity Simonett's response – expressed in ripe Australian.

In fact, imagination often is the limiting barrier, and new suggestions sometimes appear so unrealistic that they are

David Simonett
*University of California,
Santa Barbara*
1926–

214

ignored by conventional bureaucratic minds. Until they are adopted, of course, and then everyone wonders what all the fuss was about. Many years ago Torsten Hägerstrand suggested that instead of collecting census data by recording it on forms (forms that took years to process, particularly in the Third World), the census-takers should punch it on fairly durable plastic sheets. These would be displayed on the roof, or on the ground just outside the home, so that either aircraft (in those days), or satellites could pick up the images, and then process the data automatically *at the location where the information was gathered*. That was years ago: today not even a trial run has been made, although it would cost a fraction of the amount wasted annually on the dozens of international conferences that solve nothing at all. In fact for many planning purposes, reasonably accurate population figures would represent a marked advance over what many countries have today, particularly as censuses in the Third World are being constantly distorted for local political purposes.

Providing accurate counts of people *in place* was precisely the aim of Waldo Tobler when he enlarged one of the early photos of the Nile delta, taken by one of the astronauts on a Gemini flight in 1965 (Figure 18.4). We know, from the work of the Swedish geographer Stig Nordbeck, that there is a rather precise regularity between the area of a town and the number of people who live there. We can see this when we plot populations against urban areas on a graph, a close relationship that is called the 'law' of allometric growth. When the carefully measured areas of towns and villages in the Nile delta were plotted against their population values obtained from the Egyptian census – a census with a sound reputation for accuracy – the same tight allometric regularity was found. Therefore, if what you really want for national planning are basic population figures accurately located, there is no need to undertake an extremely costly census, so using scarce national resources that could be used for other development projects. All you have to do is count your urban land use pixels, multiply by a number that is little more than the slope of your allometric line, and you have a 98 per cent accurate estimate, one that is probably better than most censuses today – including those in economically advanced countries. Of course, this has never been done: good geographic ideas like these never get through the national and international bureaucracies, perhaps because too many of the bureaucrats are attending international conferences on 'Towards a Coordination of National Census Taking: Whither Tomorrow?' It is much more fun talking about it in Nairobi

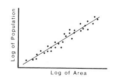

Stig Nordbeck
University of Lund
1929–

215

Figure 18.4: A photograph of the Nile delta taken by an astronaut on a Gemini flight in 1965, and enlarged by Waldo Tobler to measure the areas of the towns and villages. These were used to estimate the populations of the urban areas.

John Estes
University of California, Santa Barbara
1939–

over a cocktail at the taxpayers' expense than actually doing something about it.

But sometimes something practical is done about it. John Estes, another geographer closely associated with remote sensing since its earliest days, has helped to pioneer in many areas of practical application, not the least the sensing of large oil spills. When oil spreads on water, either from a broken tanker or an oil well leaking, the reflectance values of the water are radically altered, not only in the highly reflecting ultra-violet, but also in the thermal (infra-red) portions of the spectrum. In this way, large oil spills can be picked up and followed, making for a more efficient cleaning-up operation. The same principles are used to detect areas of cold upwelling water, for example along the coasts of Taiwan and Peru. These nearly always disclose good fishing grounds as the cold surging waters bring rich nutrients to the surface for the fish.

This now taken-for-granted ability to back away from our planetary home has allowed us to see so many things we could not see before, even though they were literally underfoot. Once again, it is that question of scale we met in Chapter 13, the same thing that Jacquetta Hawkes was implying when she wrote

216

'From too close looking follows loss of sight'. In Argentina, for example, conventional geological surveys in the northwest had mapped large areas in great detail, but they had totally missed seeing the Cerro Galan caldera, one of the largest resurgent calderas on earth. These are huge volcanic features stemming from explosions almost beyond our imagination. The newest and largest, for example, the Toba caldera on Sumatra, threw out so much material that deposits 10 centimetres thick appear in cores taken from the seabed of the Indian Ocean 2,000 kilometres away. Yet the tiny human ants with their geological hammers sometimes fail to see them.

Even large features go undetected by the ants, although they are recognized with joyful surprise once the satellite image has given them a new perspective. There is a rule that in the northern hemisphere winds tend to be deflected to the right by the rotation of the earth, a deflecting force called the Coriolis Effect. This means that winds blowing from high- to low-pressure zones do not move straight down the pressure gradient, but take on a clockwise rotation as well (or anticlockwise in the southern hemisphere). One of the most stable high-pressure areas on earth lies over the Sahara, and during the millennia of desiccation the winds have etched their clockwise pattern into the landscape of drifting sands. Today, the gross patterns in the dunes, sometimes hundreds of kilometres long, testify to the consistency with which the winds have been blowing.

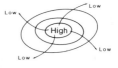

Even smaller features, like meteorite impact craters, go undetected by ground surveys. It was only when the crew members of the Apollo satellite were over northwest Brazil that the rings of three meteor craters were discovered, one of them 4 kilometres in diameter. Ground surveys later confirmed the distinctive fracture patterns that indicate the enormous shockwaves produced by such extraterrestrial impacts.

Sometimes, of course, human beings leave their own 'undetected' prints on the earth's surface, both large and small. Archaeologists have long known that if you take air photos at sunrise or sunset, the slanting sunlight will throw into relief features undetectable on the ground. Colour photos also pick out features wherever the soil has been disturbed, particularly if the area is now being used for fields of grain whose uniformities tend to highlight any slight contrast in colour. Perhaps one of the most moving pictures is an area of eastern France, where the great zig-zagging slashes of the First World War trenches are superimposed upon the roundels of burial sites from the Bronze Age. Now the men of two worlds, thousands of years apart, sleep together, only the colours of the waving wheat testifying to their final resting places (Figure 18.5). Today archaeologists in Central America are exploring the use of satellite imagery to extend our knowledge of past civilizations. Many of these archaeological features are buried beneath the vegetation and soil of dense jungles, but as the pixels get smaller and the resolution of our sensing increases, we may be able to see the small signs that human beings unwittingly left behind.

Figure 18.5: A photo taken over eastern France showing the zig zag slashes of the trenches from the First World War and the round burial sites of people from the Bronze Age. Both are highlighted by distinctive changes in colour in the large wheat fields.

One of the most exciting technical developments that is helping geographers and others see things that are hidden away is SLAR, or sidelooking airborne radar. Instead of relying on what we might loosely call 'natural' radiation, an intense beam of microwave energy (beyond the infra-red) is sent out by the aircraft or satellite, bounced off the object to be sensed, and then recorded upon its return. The advantage is that such radar waves 'cut through' soft stuff like vegetation and soils, and give us an image that is literally 'rock hard', as though all the surface

218

materials had been stripped away, leaving only the bare rock bones exposed. You can imagine what this means in areas of dense tropical jungle or forest. On the island of Flores in the Indian Ocean, for example (Figure 18.6) the SLAR image looks

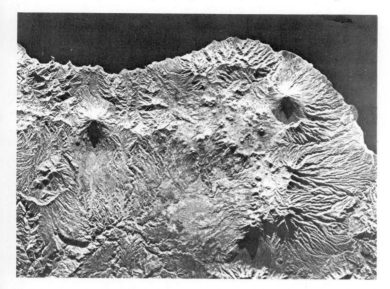

Figure 18.6: A sidelooking airborne radar image taken through total cloud cover of the island of Flores in the Indian Ocean. The radar was beamed from the north to 'highlight' the 3-dimensional effect, and it cuts through all vegetation and soil to expose the underlying shape of the surface.

almost like one of those three-dimensional relief maps that schoolchildren sometimes build up layer by careful layer. The marvellous effect of highlighted relief comes from the fact that the radar wave was originally beamed from the north (the top of the picture) to 'illuminate' the slopes from that direction, so throwing into shadow the opposite sides. If you turn this book upside-down, those volcanic peaks look like deep holes in the ground. Strange, isn't it?

But radar images are not just taken because they are dramatic or pretty, and usually there are very serious purposes lying behind such expensive pictures. For example, the Indonesian government has been trying to establish what it calls its Transmigration Programme, opening up areas of jungle to agriculture to produce more food, and to take the population pressure off the extremely densely populated areas of Java and Sumatra. The problem is that topographic maps in areas of dense jungle are often inaccurate, and sometimes survey parties on foot have missed whole mountains. Heavy equipment is difficult to move in unsurveyed terrain, and areas may be chosen for development that are poorly drained or subject to flooding in the wetter seasons. Today, Kalimantan (southern Borneo) is being sensed by long-wave radar that cuts like a razor through obscuring vegetation, and discloses the fine-grained drainage features that are so critical to agricultural planning and development. It does not miss mountains, either! Similar successes have been achieved in Panama and the Amazon Basin, where new and unsuspected tributaries have been found. The radar beams even cut through thick layers of sand. In Iraq, Saudi Arabia and Egypt the old water courses from more humid times are disclosed, as though the radar beam were a huge whisk broom sweeping the covering sands of centuries away. These former drainage patterns provide clues for providing fresh water sources today.

Perhaps the most dramatic radar pictures have come from our new-found ability to back away and see the earth as a whole. In 1978, a satellite called Seasat bounced radar waves off the surface of the sea, recording the height of the surface to an accuracy of a few centimetres. Since the height of the ocean's surface is directly related to variations in the earth's gravitational field, it was raised up to 4 metres over underwater seamounts, and depressed by the same amount over the deep sea trenches (Figure 18.7). In two or three *weeks*, Seasat was

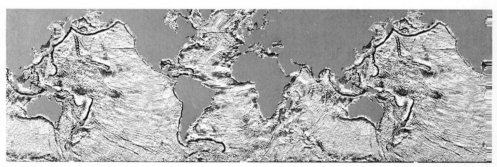

Figure 18.7: Variations in the sea surface, and so the underlying sea bed and force of gravity, taken by SEASAT. The great plates and ridges can be easily seen, and in a few weeks the satellite collected more data than had been obtained during the past 60 years.

able to record more information much more accurately than during the previous sixty *years* of conventional approaches. Geographers and geologists have found whole mountains under water never detected before, and have outlined the great fracture zones where the huge plates of the earth meet. If we had had such pictures thirty years ago, all the intellectual energy used to denigrate the theory of continental drift might have been used for more constructive thinking.

For the geographer, the age of not-so-remote sensing has long arrived, and today the basic skills of transforming numbers on magnetic tape to multicoloured images disclosing the wonders of our home are taught in most departments of geography. Slowly – much too slowly – the security classifications are being lifted, and sensing for civilian purposes is coming more and more to the fore. New agricultural schemes, regional planning, crop estimations, energy saving . . . the list of uses is growing every day. That little blue speck in the infinite darkness is being scrutinized ever more carefully. With luck and intelligence, we will see enough to save our home rather than destroy it.

Part VI
Teaching and helping

Geography and medicine: an old partnership

19

One of the most exciting things about the new developments in geography is that they have produced a great deal of cooperation with people working in other fields. A particularly fruitful partnership has been formed with medicine, particularly the subfield known as epidemiology – the study of how diseases spread. Since geographers have played such an important part in the development of the theory of spatial diffusion (Chapter 12), this is hardly surprising, but the contacts between geography and medicine go back to earlier days, and to simpler, but still extremely important and provocative, forms of analysis. Not the least of these is the straightforward plotting and careful examination of data on maps. It almost seems that when doctors become geographers, or geographers get hold of medical data, previously unasked questions come out of concealment and bubble to the surface. Perhaps the most famous example in this old partnership is the work of Dr John Snow in the middle of the nineteenth century, who backed up his hunch that cholera was spread by infected water supplies by making a map of victims in a part of nineteenth-century London. This was during the days when medical opinion ascribed the disease to 'miasmas', and other emanations from the swamps and mudflats of the Thames. His map showed a nebula-like cluster centred on a public water pump, and when the pump was closed down, the cholera stopped.

This same tradition of keeping a *geographic* account (location) of a disease, as well as an *historical* account (time of infection), continues sporadically to this day, all the way from official international bodies collating worldwide information, to the devoted efforts of single doctors working under the most difficult conditions. Smallpox was tracked by the World Health Organization (WHO) year after year, until the last pockets in Ethiopia and Somalia were squeezed out of existence. *Where*

225

it was became a crucial part of the campaign that assigned vaccination teams to *places*. At the individual level, clusters of Burkitt's sarcomas in women were meticulously recorded on maps in an isolated rural clinic in northern Uganda. These eventually led to new lines of cancer research when it was realized that nearly all the cases lay in the valleys of endemic malaria, while the women living on the ridges were virtually free of cancer.

Yet even at these relatively simple levels, geographers and doctors often have to work under three rather severe handicaps. The first we have met many times before, and it is simply that old problem of getting people to record locations – really just a matter of attaching x and y coordinates, or latitude and longtitude, to any person's medical history. It seems such an obvious and simple matter, but few people besides geographers are really taught to 'think spatially' and it is too often forgotten. To put it bluntly, you cannot map and follow the spread of disease if you do not know where the cases are. The second problem is one of diagnosis; you have to be able to identify the disease, and you have to record it and report it in some way. Two more naively obvious points, but sometimes difficult to carry out. You would think that something as horrible as bubonic plague would be easily diagnosed by any medical student, let alone an experienced doctor, but the truth of the matter is that most medical people have never seen a case of this now-rare disease. Few doctors in medically advanced societies ever think 'plague' – it literally never 'comes to mind' – so when cases suddenly appear in New Mexico (where it happens to be endemic in the wild animal population), it is not immediately diagnosed. Other diseases that the layman lumps under such names as cholera or 'flu are actually much more complex, and may require extremely sophisticated diagnostic tests and equipment to identify. Is the cholera the normal Ganges sort, or is it the notorious El Tor strain that lay quiescent in the Celebes for over half a century before it exploded outwards in the 1960s and early 1970s (Figure 19.1)? Is the 'flu any one of several Hong Kong varieties, or Victoria Type A, or Type B? And after a disease has been positively identified, is it reported? The World Health Organization, with its Center for Disease Control at Atlanta, asks all nations to report certain diseases. Some do their best to comply with such international efforts, but others do not – partly because there seems to be a sense of shame in reporting certain 'uncivilized' diseases.

This same sort of sensitivity also makes for grave difficulties

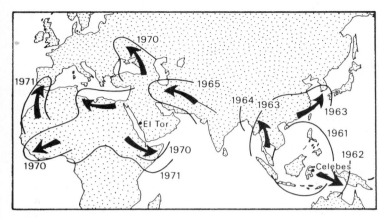

Figure 19.1: The diffusion through Asia, Africa, and parts of Europe, of the notorious El Tor strain of cholera from its source region in the Celebes. Apart from a jump to West Africa in 1970, it shows a strong effect of spatial contagion.

at the individual level. In many countries, venereal diseases are considered shameful, so people will avoid going for medical advice. It is virtually impossible to study the epidemiology of syphilis in the United States, even though it is a highly dangerous and reportable disease. The same problems hold true in other areas of medicine; for example, birth control. In a small town in Pennsylvania it proved impossible to study the diffusion of planned parenthood because certain religious teachings placed an enormous burden of guilt on some women who did not want anyone they knew to see them go into the newly opened clinic. The issue was so delicate that the medical board reluctantly forbade a study in which they had a deep interest. Obviously, there are questions of confidentiality in medicine, but equally we cannot learn about various aspects of public medicine without data or responsible research. Location, diagnosis, reporting and confidentiality – you wonder sometimes how geographers and doctors make any sense out of it at all.

Despite the difficulties, some extremely valuable work has been done, and it seems to be growing day by day as the geographic perspective and spatial awareness of people slowly rise. One of the most exciting examples comes from China, where the Academy of Medical Science sponsored an extraordinarily detailed atlas of various sorts of cancers. Some areas, and even some individual communes, showed strikingly higher rates than the surrounding regions, and these maps pinpointed areas for further intense research and treatment. High rates of liver cancer were found in areas where a corn and rice fungus produced a deadly toxin, and this is now being controlled by proper food processing. In the northeast, women were found to have much higher rates of lung cancer, although they smoked

227

far less than the men. When their daily lifestyles were examined, it turned out that they used the low-grade coal in the region for cooking and heating, and spent much of the day in unventilated areas inhaling the smoke. Perhaps the most dramatic discovery was the unusually high rates of cancer of the œsophagus among certain communes in Henan, where pickled vegetables, a local delicacy, provided a culture for a carcinomic mould. Paramedics – the barefoot doctors – were taught how to diagnose the cancer by asking people to swallow balloons, which were then gently inflated so that they pressed against the inside of the throat when they were withdrawn. A simple microscopic examination of the sputum easily showed up cancer cells if they were present, and a simple operation was able to save many lives.

A rather similar atlas of cancer was also made for England and Wales (Figure 19.2), and the map of male cancers is almost

Figure 19.2: Deaths from cancer among males in England and Wales, 1968–1978. The map is an almost exact replica of the urban areas.

a perfect match of the urban areas, raising some broad questions about the sort of environments these form for human beings. Underneath these broad scale patterns, however, there may be many specific causes, and it is the questions that these more detailed patterns raise that can be investigated more carefully. Unusually high rates of nasal cancers seem to be associated with the tailoring of women's garments (would a check of such rates in the 'garment district' in New York City bear out such an association?), as well as with leather goods, furniture and upholstery. What solvents are used in such manufacturing processes? Similarly bladder cancers seem to be associated with the dye stuff and rubber industries. Of course, such maps can only raise questions, not answer them, but good questions are always the starting points of scientific inquiry, and *how* medical data are mapped may well determine whether a question is raised or not.

The question of how medical data appear on maps is an important one, and if individual cases of diseases do have geographic coordinates attached to them as part of the medical history there is no problem. Maps at any convenient scale can be compiled to highlight those always questionable clusters. After all, as human beings we are pattern seekers, and our eyes are drawn to unusual distributions or blobs on the map. At Seveso, Italy, 16 days after a chemical explosion had deposited white flakes of dioxin downwind (children actually played in the 'snow'!), animals and plants were keeling over, a score of people were in hospitals with severe skin lesions and vomiting, and the chloracne cases virtually outlined the fallout downwind (Figure 19.3). Such a map, constantly revised day

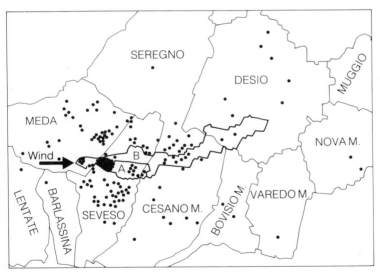

Figure 19.3: The ICMESA factory (arrow) explosion at Seveso in northern Italy that scattered quantities of dioxin downwind. Each dot represents a severe chloracne case. Area A was evacuated, and B was 'cleaned up', but in the rest of the region no action at all was undertaken.

by day in the local newspapers, might have done much to alert people to move. As it was, the health minister of the region said on TV that 'everything was under control', and a group of 'scientists' in nearby Milan said they would be quite willing to live in the area closest to the plant explosion. Needless to say, their bluff was never called, and none of them volunteered.

30 times the national average

4 times the national average

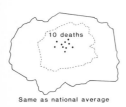

Same as national average

Unfortunately, most medical data are not located so precisely, but are lumped together, and mapped in some convenient census or political unit that may have no bearing on medical problems whatsoever. In this way, an interesting cluster in a small area can be swamped by being included in a much larger reporting unit. It was by disentangling these effects of spatial agglomeration that the very high rates of leukemia in young people around the nuclear reprocessing plant at Sellafield were discovered. Figures reported in local health statistics, using the 'official' units, showed only slightly higher than average rates, but when the data were remapped the rates in two areas were 12 and 5 times the national average. The problem was that the radioactive discharges from the plant into the Irish Sea formed part of the silt on the mud flats, and when these dried out radioactive plutonium, americium, caesium and ruthenium were blown into homes to be sucked up into vacuum cleaners and human lungs. Even pico-scopic quantities of these elements may be deadly to young children, because the stomachs of babies in their first year of life absorb ten times the amounts of adults. So *how* you map such data is a critical geographic question that determines whether other medical questions are asked or not.

Nor are such geomedical questions confined to physical symptoms: we seem to live in ever more stressful environments, and sometimes we can cope no longer and go over the brink. When entrance rates into psychiatric wards of hospitals around Heathrow Airport (the suburbs of West London) were examined and mapped, startlingly high rates far above the national average were discovered. People could no longer cope with the almost constant scream of jets taking off and landing, and the same problem arose in Los Angeles – where a wide swath of residential area was cleared at the cost of hundreds of millions of dollars in compensation. Great resentment was caused in France and Britain when New York City would not allow the supersonic Concorde to land at first, but a careful study by the National Academy of Sciences had already demonstrated intolerable levels of noise around Kennedy and La Guardia airports. At a number of schools, teachers were unable to communicate with their pupils for ten minutes out of every hour even when they were shouting at the top of their voices.

Environmental disturbances may be silent too, and over the last eighty years modern power and communication systems have totally altered the electromagnetic environment in which we live. The effects of microwave radiation have yet to be investigated properly, but during the Falklands War the components of the light cassette recorders used by reporters were literally fried on the ships generating radar beams. The question is what does such radiation do to people? What we do know is that other electromagnetic fields, generated by ordinary power lines, are closely associated with depressive mental illness and increased rates of suicide. One very carefully designed study in the Midlands of England located the residencies of nearly 600 suicides, and the same number of ordinary people as a control group. When the electromagnetic fields were measured, it was clear that the suicide group had been subject to significantly greater electromagnetic radiation, confirming the effects discovered in other studies on childhood cancer, reduced sperm counts, and radically altered imbalances in the proportion of males to females born to high-voltage workers. Since we are all subject to some levels of such radiation (even from the electrical wiring in our homes), we have here a vital area of investigation in which locations, distances and distance decay effects are an integral part of the research.

But research today in what we might call 'geomedicine' goes far beyond the static plotting of cases on maps. Diseases are anything but geographically static, even if our maps do tend to show us frozen 'time slices' or geographic snapshots of an ongoing, and often highly dynamic, process. Right back in the old Greek roots of the word *epidemic* we find the sense of something moving through a population of people (*demos*), spreading outwards (*epi-*) from a central point, like a brush or forest fire blazing its way across the landscape. But we have to be careful here, because this 'wine-stain-on-the-tablecloth' image is a seductive one, and we know from Chapter 12 that it implies a process of *contagious* diffusion – a term that geographers have borrowed straight from medicine. We can certainly find historical cases, where thousands, and even millions, of transmissions of a particular disease at the micro, or individual level show themselves as great waves of contagious diffusion at the macro level. Working with colleagues in history, Gerald Pyle has examined some of the great influenza pandemics moving through Europe over and past 400 years (Figures 19.4 and 19.5), and it is clear that when the friction of distance was high (1781–2), the disease rolled from east to west like a great comber breaking and dissipating itself on the beach. The same pattern appears 100 years later (1889–90), suggesting rather stable historical connections, but notice that it only takes one-half the time now to spread from Russia to Portugal. Something has happened to that friction of distance in the meantime, for now the railways begin to structure the space we call Europe.

Gerald Pyle
University of North Carolina at Charlotte
1937–

231

Figure 19.4: The
diffusion of influenza in
Europe during the
1781–82 pandemic.
Arriving from the
Russian Empire, it
took nearly eight
months to spread to
the Iberian Peninsula.

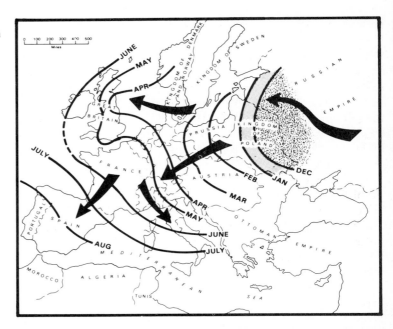

Figure 19.5: The
diffusion of influenza in
Europe during the
1889–90 pandemic.
With a greatly lowered
friction of distance one
hundred years later, it
took only four months
to sweep through the
continent.

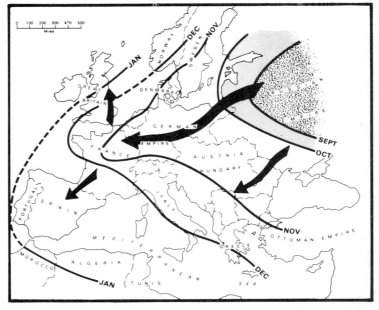

It is the *structure* of geographic space, the changing intensities
of connections, that become the crucial aspect of any study of
an epidemic, and this is where the expertise of the geographer
comes into full play in this boundary zone between geography

232

and medicine. Other studies by Pyle of the great cholera epidemics in the United States during the nineteenth century have shown how the course of the same disease is altered by changes in the connections and structure of the space. In 1832 (canal and stagecoach days), the disease seemed to be highly controlled by distance – the classic contagious diffusion – while the last epidemic of 1866 (with the railways beginning to change connections), appeared to jump much more down the urban hierarchy. But even in this rather rare case, where we have the opportunity to examine the course of the same disease over a time of rapid change in transport technology, the issues are not simple, and the effects of distance and population size are difficult to disentangle. Approaching this problem of sorting out the convoluted effects of contagious and hierarchical diffusion, we come face to face again with that same geographic headache we met in Chapter 11 – the intertwined and inter-related effects of distance and map pattern. Do you see now why I led you through an apparently esoteric discussion? It is this sort of theoretical and technical difficulty that lies at the heart of our ability to understand, and to come to grips with, every epidemic we might try to control.

Peter Haggett
University of Bristol
1933–

As human beings, we cannot just study epidemics in a detached and unconcerned way: underneath those maps we construct and interpret there is a deep human concern to do something about the epidemics displayed, a desire to intervene in the spread of the diseases if we possibly can, and stop the human suffering they represent. The medical person brings an irreplaceable skill for prophylactic intervention, diagnosis and treatment, but an *understanding* of why and how a disease spreads rests squarely in the modern geographic realm with its concern for spatial structure. A disease is a traffic transmitted on a structure: if you do not understand the deeper structure, you will never understand the transmission.

Andrew Cliff
University of Cambridge
1943–

Nowhere is this intimate concern for spatial structure and dynamics seen more clearly than in a study of measles that has quickly become a classic. Undertaken by Peter Haggett, Andrew Cliff and Keith Ord, with a colleague in historical illustration, the research focused upon Iceland; in part because of its isolation, in part because of the meticulous medical records that are so typical of the Scandinavian tradition, and in part because every town, including the capital Reykjavik, was below the 'endemic threshold'. This means that the populations were small enough for measles to die out completely (Figure 19.6a), so that every new epidemic meant a reintroduction of the disease from abroad – usually from boats arriving

Keith Ord
The Pennsylvania State
University
1942–

Figure 19.6: In a small, fairly isolated country like Iceland (a), measles displays its distinctive periodicity, but because all towns are below the 'endemic threshold' the disease can die out completely. In contrast, some of the towns in better-connected Denmark (b) are large enough so that measles never dies away completely.

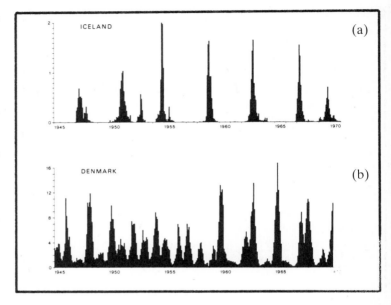

from Denmark or Norway. This is in marked contrast to Denmark (Figure 19.6b), where the disease never completely dies away, but is always 'somewhere' in the population waiting to break out again. In a sense, Iceland represents a little geographic laboratory, a small and reasonably isolated structure that might be simple enough to understand, while measles is an 'ideal' disease to study – if we can bring ourselves to use such an adjective for a disease that sometimes causes death among small children. What I mean, of course, is that it is easily diagnosed, it is reportable, and it has a distinctive cyclical rhythm, due to the fact that the disease confers immunity once a child has it. Once all the children in a village have had measles, it takes about four to five years for another group to come along who are susceptible.

If we look at lots of measles epidemics in Iceland, and generalize some of the main features, we can see a sort of mixed hierarchical and contagious process going on at various scales (Figure 19.7). About five years after the last epidemic, perhaps a boat arrives in Reykjavik from Bergen, Norway, with people with measles aboard. They escape the quarantine, perhaps by claiming they have had measles before, or, more likely, because no one is aware that they have the disease when they land. They either travel to their homes elsewhere in Iceland, or give it to a visitor in Reykjavik who takes it back to a smaller town. At the local level, it is almost certainly the

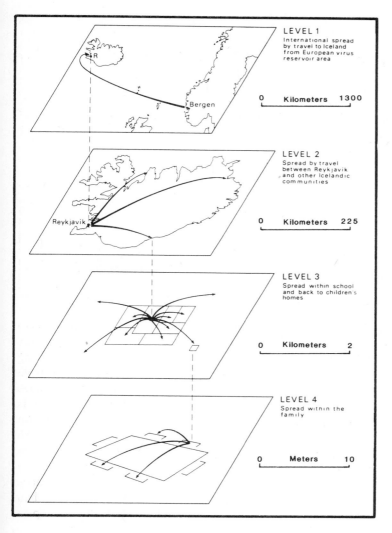

LEVEL 1
International spread
by travel to Iceland
from European virus
reservoir area

0   Kilometers   1300

LEVEL 2
Spread by travel
between Reykjavik
and other Icelandic
communities

0   Kilometers   225

LEVEL 3
Spread within school
and back to children's
homes

0   Kilometers   2

LEVEL 4
Spread within the
family

0   Meters   10

Figure 19.7: The mixed contagious and hierarchical diffusion effects in Iceland operating at various geographic scales. A boat arrives at Reykjavik carrying measles from Bergen in Norway (Level 1). Within Iceland, people carry the disease to the smaller towns and communities (Level 2), where the children in school (Level 3) finally carry it home to small brothers and sisters (Level 4).

children who spread it quickly, particularly in the schools, and they then take the disease home to give it to younger brothers and sisters. An actual epidemic (1946–7) spread more or less in this way (Figure 19.8a–d), arriving at Reykjavik (a), spreading to three smaller towns on the other side of the island (b), hopping through the small towns in the more densely settled southwest (c), and finally spreading to nearly every little town and village (d) before it died out.

On the surface, it seems a fairly simple task to model this diffusion process, and so be in a better position to predict where it will strike next. Such information might be critical

235

Figure 19.8:
Geographic
'snapshots' of the
spread of measles in
Iceland during the
1946–47 epidemic.
From Reykjavik (a), it
spreads to three towns
on the opposite side of
the country (b), then
to the densely-settled
southwest (c), and
finally all over the
island (d).

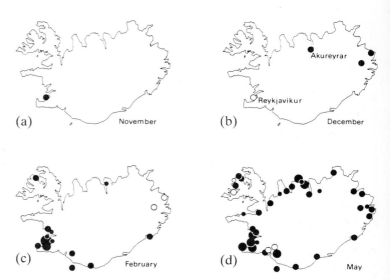

if we wanted to intervene in an epidemic with a vaccination
programme. Unfortunately, what this study showed, in a
meticulous step-by-step attempt to construct more and more
complex models, was the difficulty of disentangling the convoluted effects of space and time, although it seemed that the
geographic pieces of such models (the spatial structure) were
more important than the exact time phases in deriving the
predictions. Epidemics are not easy to understand, and in the
case of measles we are discussing the sixth largest killer in the
world, a sombre fact that few people appreciate in countries
where children are routinely vaccinated today. Yet it is the fact
that measles *can* die out completely in small communities, and
can be vaccinated against, that holds out our greatest hope.
Even in the United States, with its large population and huge
cities ensuring that measles will be endemic, the vaccination
programme has had dramatic effects (Figure 19.9). Since 1965,
measles epidemics have almost been wiped out. In 1974,

Figure 19.9: The
regular rhythm of
measles epidemics in
the United States, but
notice the severe
dampening of the
disease's incidence
after the 1965
vaccination programme
started.

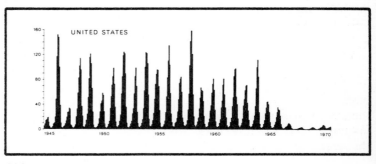

smallpox was fifth on the list of the world killers; today it has disappeared, the first disease to be totally eradicated. Perhaps with the same efforts, measles can go the same way.

Unfortunately, for other diseases, even the most common ones, it is difficult to hold out such hopes, and the medical-geographic partnership has still a long way to go. Influenza is a recurrent disease everyone is familiar with, and its position on the list of leading causes of death (eighteenth in 1974), not only fluctuates as each pandemic roars around the world, but it is seriously underestimated by poor diagnosis, poor reporting, and the fact that it often contributes to other 'causes', such as pneumonia, bronchitis, heart failure and so on. Time and again, we have tried to come to the most elementary understanding of this disease, let alone attempt to model it and predict it. Every time it seems to slip through our fingers. It changes its coat with startling rapidity, so diagnosis is difficult, and it is a plain and simple fact that we do not know very much about it. In the northern hemisphere it appears with almost clocklike regularity around November, rises to a peak in January or February, and starts to wane away in the spring. Where does it go to? Where does it hide? We can diagnose and trace new strains (and so have about six months' lead time to prepare vaccines), and we can follow them more or less around the world when they first appear. But then older forms of 'flu, apparently lying dormant during the summer months, take off again, and we are lucky if we get 'flu diagnosed and reported properly at all.

In the United States, the Center for Disease Control publishes weekly 'deaths from pulmonary causes' for 131 cities, and these crude figures might be used as surrogate measures for influenza, since you can 'see' a 'flu epidemic as a distinctive bulge in a time series of a particular city. You would think that since influenza is carried and spread by people (except for a few pigs who can also be hosts for the disease), that if we know how people move around the United States we could also model the diffusion of influenza. After all, even simple gravity models predict air traffic quite well, so we ought to be able to 'inject' 'flu into the structure of transportation and watch it diffuse in fairly predictable ways. No such luck: no matter how we massage the huge quantities of data, we are no closer to understanding the course of an influenza epidemic than we were before. Too many things, including different strains and the seasonal appearances and disappearances, combine in unknown ways to let us tease apart the varying effects. What we need are quick, cheap and simple diagnostic tests for

237

particular 'flu viruses, and a much tighter reporting system that gets the data to the Center for Disease Control in an accurate form. Only then will we be able to get a proper grip on the process. In the Soviet Union much greater progress is reported, and 'flu predictions are apparently used to allocate vaccines, first to critical occupations such as firemen, police and teachers, and then to other people. It is reported that such intervention saves millions of people-hours every year, and the billions of roubles that such hours represent in the form of production – let alone the widespread human misery we all know, and the increased medical complications for the very young and the elderly.

Medicine and geography: an old partnership that was almost dissolved, but one that is being forged once again to the benefit of both fields, and to the people of the larger society that both fields serve. For the driving force is to help, and the medical concern of the modern geographer ranges widely, and includes people whose handicaps make it difficult for them to fit in with the swirl and bustle of modern life. It is time we took a look at these closely allied traditions of contemporary geographic research.

Mental maps and geographic prisons

20

If we are lucky enough to be endowed with some intelligence, and a reasonable amount of fortitude, we can usually make some sort of sense out of the strange and wonderful world we find ourselves in. Information about our world comes to us in various ways to be chosen, memorized, forgotten, structured, interpreted and valued, and we use it constantly to steer ourselves through the rough waves that buffet us from time to time on the sea of life.

That image of steering through a calm or turbulent sea can be thought about in a number of different ways. We may find ourselves with a difficult problem or decision on our hands, and sometimes we can almost feel ourselves turning this way and that, following one train of thought that leads to an unacceptable end, going back to the beginning to start again, trying another *tack* – so we must be in a boat after all! Wrestling with difficulties, trying to 'find our way', is a common experience for all of us, and from time to time we face the problem of way-finding in a direct physical sense. Which ski track in the deep woods will take us back to the village? Which street leads to rue Quincampoix near the Beauborg? Is 54th Street this way, or in exactly the opposite direction? I can't recognize a thing . . . where am I?

Making sense out of information, recognizing key marks, and finding our way can be difficult even when we have reasonable endowments of intelligence, commonsense and fortitude, because sometimes our mental images seem to have little congruence with something we are pleased to call 'reality'. This is not the place to get into a long and philosophically deep discussion about what we *really* mean by *reality*, but let us recognize that most people seem to make reasonable sense out of their world most of the time, and they find their way about for not only the daily humdrum tasks of shopping, but also the long trips to other places. Yet there are many who have great difficulty in making sense out of the world, people whose

Reginald Golledge
*University of California,
Santa Barbara*
1937–

Julian Wolpert
Princeton University
1932–

Eileen Wolpert
Independent Scholar
1934–

'mental maps' do not fit very well, and who are gravely handicapped when it comes to those basic 'sense-making' endowments the rest of us usually take for granted. Until, perhaps, a severe accident, or a debilitating disease, reduces our own ability to make sense and steer. Then we too may need some navigational help.

The idea of a mental map – perhaps a cognitive or environmental image – is something that is often of deep concern to geographers as they try to come to a better understanding of the way human beings behave in the space(s) around them. Reginald Golledge has been at the forefront of such research, often working with colleagues in psychology and mental health, and his concern for the mentally handicapped, and the difficulties they experience in coping with their world(s), is deeply shared by Julian Wolpert. Attitudes to mentally handicapped and retarded people are changing rapidly today, and the feeling is growing that they have the right to live in a society whose levels of compassion and understanding can be enhanced by humane education. Working with his wife Eileen, and other colleagues, Wolpert has focused attention on the problems of deinstitutionalizing people without providing proper social support, pointing to the new ghettos of rundown boarding houses and seedy hotels that are appearing close to downtown in a number of American cities. They deal with this tragic theme in a variety of ways, ranging from descriptions of individual and community responses to mentally handicapped people (including the considerable opposition to the siting of drug and alcoholic centres, halfway houses and boarding and care homes), to highly formal mathematical analyses of the optimum way to allocate scarce (?) resources to handle the hundreds of thousands of newly released people.

A basic question is whether such people can even find their way about the urban environment in which they find themselves, for without such skills even the simplest daily tasks become impossible. In Santa Barbara, California, for example, Golledge compared the ability of ordinary people to recognize landmarks and navigate 'the maze' with those who were mentally handicapped in varying degrees. In an earlier study, in Columbus, Ohio, a few people had excellent abilities (Figure 20.1a) while some made some mistakes, more than you or I would perhaps (Figure 20.1b), but still were able to cope with basic daily tasks satisfactorily. But another poor soul had virtually a random mental map (Figure 20.1c), a collection of things without any connections, and therefore without any meaningful structure to it.

240

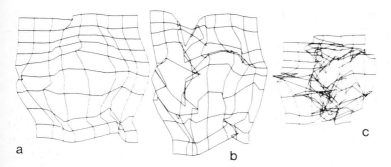

Figure 20.1: Some people have excellent 'mental maps' (a) which correspond almost exactly with the surveyors. Others (b) make some mistakes, and produce a few distortions, while the mental images of some people (c) bear little resemblance to 'reality'.

a

b

c

Whether mentally handicapped or not, we all hold certain images of places around us, and, more importantly, we all have our own values, likes and dislikes, so we tend to place somewhat different meanings on them. Even the way we judge distances can be quite unwittingly shaped by the images we have acquired. During the Vietnam war, for example, American university students tended to overestimate the distance to Hanoi, and underestimate the distance to Saigon, even as they pulled Europe towards them and pushed Africa away. A decade later, Dutch students at Amsterdam also distorted the measured space of the cartographer (Figure 20.2), stretching

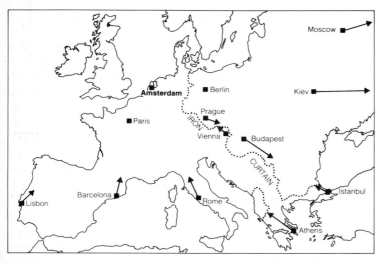

Figure 20.2: Students in Amsterdam estimate the distances to Paris and Berlin quite well, but tend to 'pull' the cities of western Europe towards them, while 'pushing' those of eastern Europe away. The mental effect of the Iron Curtain is clearly seen, and appears as a consistent effect each time the experiment is repeated.

and squeezing the familiar map of Europe as though it had been drawn on a rubber sheet. Distances from Amsterdam to Paris and Berlin are estimated fairly well, but notice how every city east of the Iron Curtain is pushed away, while even Istanbul is drawn closer to them. Athens appears much closer to them, and perhaps it really is – in 'reality'. So *now* what is reality?

The Greek world holds a very special place in the Western imagination, and perhaps 'in reality', not simply in old-fashioned geographic space, it appears quite close and a part of things. It turns out that these 'distortions' are highly consistent year after year, and we have quite a lot of evidence that mental images, shared by similar groups of people, tend to be extremely stable.

We can see from this simple, but rather dramatic example of space stretching on either side of a political dislocation like the Iron Curtain that considerable differences in perception may be caused by barriers – although we must be very careful not to jump to conclusions too quickly. It is true that jagged mountain chains have historically divided and isolated people, even the people of one valley from another, but what about a mountainous, but highly unified country like Switzerland? It is also true that water can divide, as the English and the French know, yet rivers, so often used as national boundaries to separate people, may actually unite. Even today, it is probably easier for a wine-grower in the foothills of the Vosges mountains in western France to understand (in all senses of the word) his fellow in the Kaiserstuhl area across the broad Rhine river in Germany, than to communicate (in all senses of the word!), with an official from Paris. How close do we have to push England towards France (or is it France towards England?) to get real communication going? Conversely, if we had a region culturally united around a river valley, and we could gradually pull the two sides apart, at what point would unity become separation?

In a less fanciful sense, this is exactly the sort of question that Gunnar Törnqvist was asking in an applied research project of the Öresund, the 'sound' that separates Denmark and Sweden, with Copenhagen and Helsingør on one side, and Malmö and Helsingborg on the other (Figure 20.3). Distances range from 3 to 15 kilometres (wider than the Rhine, but less than the Channel), and historically the whole region was united under the Danish crown. Only when Skåne, on the Swedish side, was liberated and restored to the fatherly fold in the seventeenth century (according to the Swedes), or torn shrieking from the bosom of Mother Denmark (according to the Danes), did the two sides start taking their separate ways. Today, there are astonishingly few contacts across the barrier sound, since Copenhagen faces west and south to Europe and a larger world of international business, while Malmö and Helsingborg look northeast towards Stockholm. These different ways of looking are reflected in all sorts of different environ-

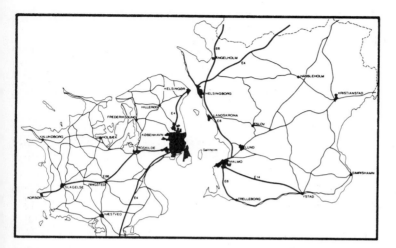

Figure 20.3: The Öresund Region, with southern Sweden on one side, and eastern Denmark on the other. Although once under the Danish crown, and 'united' today by the 'Sound', the busy channel to and from the Baltic, the differences between the two are marked. Will these change if a bridge or tunnel is built to supplement the ferries and hydrofoils?

mental images which are shaped, quite naturally, by varying experiences and streams of information. On the Swedish side, the cognitive images of high school students at Lund are very strong in Skåne (Figure 20.4); where the values 0–3 mean few

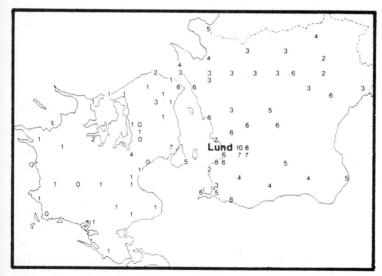

Figure 20.4: The 'knowledge map' of highschool students in Lund on the Swedish side of the Öresund. Values of 0–3 mean little, if any, familiarity; 4–7 mean some familiarity, but help needed to get there; 8–10 mean most people know the place, and have even spent some time there.

have ever heard of places there, values around 5 mean they 'sort of' know but could not get there without help, while the highest values mean almost everyone knows the place, usually from spending a considerable amount of time there. For Copenhagen students on the other side (Figure 20.5), it is exactly the opposite, with a rich mental map in Denmark, and practically no knowledge of Sweden outside of the few coastal towns. I

243

Figure 20.5: The 'knowledge map' of highschool students in Copenhagen on the Danish side of the Öresund. Knowledge of Danish places is quite good; but knowledge of Swedish places, even those quite nearby, is poor.

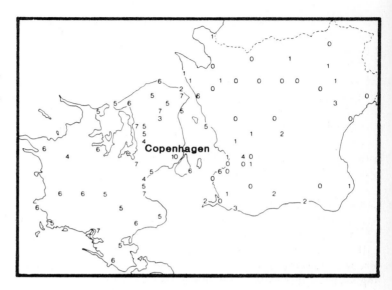

can recall a biannual visit by geography students from Copenhagen to Lund, and for some of the Danish students it was the first time they had visited the town only 35 kilometres away. And although the Swedish and Danish languages are so close that they can be read with ease by either nationality, many of the conversations between students took place in English.

These same sorts of marked disparities in mental images were found in other groups, notably in high-level businessmen and official 'decision-makers'. Great uniformity of viewpoint was found within the two groups looking at, evaluating and judging the particular conditions on their own sides of the Öresund, probably because they had had the same sorts of education, read the same journals and newspapers, listened to the same television programmes, and reinforced each other's opinions with their frequent contacts. Once again, a question of information. But as soon as they were asked to comment about conditions on the other side, all the usual stereotypes and clichés started to appear. Swedish business was more efficient, said the Swedes. Oh no, Danish business is, by far, said the Danes. Swedish business is more controlled and regulated, said the Danes. Nej, quite the reverse, said the Swedes. Only when both groups went up to the international level, and looked at Europe as a whole, did they see eye to eye almost perfectly. These mental images of people who have the capacity to influence and make important decisions – locating centres of employment, regulating against pollution, planning of all sorts – take on increased importance when there is the possibility of radically restructuring a region, and, as we know now, this always means changing the connections. The large research programme, of which the work on environmental images was but a part, was concerned with the changes that might come about when new bridges or tunnels were built from Malmö to Copenhagen, or Helsingør to Helsingborg. Would Copenhagen 'reach out' and grasp Malmö under its umbrella of influence, turning her away from Stockholm? Or would such increased connections mean greater contacts, more information on both sides, and so greater unity of perception about the region as a whole?

Such questions are difficult to answer, because even when quite similar groups of people are close together in geographic space, they may still be exposed to quite different streams of information, and translate the information into startlingly different evaluations and images. In Montreal, close to the French-speaking Quebec and English-speaking Ontario border, high school students at Pointe Claire and Pointe aux Trembles were only a few kilometres apart. But their schools were English and French respectively, and their mental maps faced in totally different directions. Shaped by French newspapers and television, visits to family in Quebec, and the general pride of being Québecois, the French-speaking students would like to live in Montreal or Quebec City, two major peaks on the preference surface (Figure 20.6), although all of the St Law-

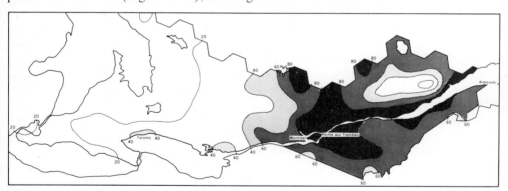

Figure 20.6: The residential desirability surface for French-speaking students at Pointe aux Trembles in Montreal. The whole surface is oriented towards Quebec and the lower St. Lawrence Valley, particularly the major French-speaking urban areas.

rence valley is highly regarded, and the perceptual gradient to Ontario is steep. As for the English-speaking students (Figure 20.7), most of them would not be seen dead in Quebec, for

245

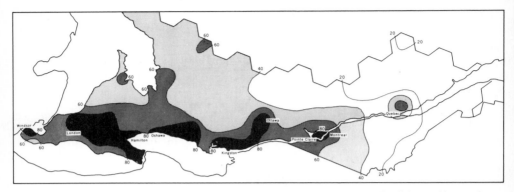

Figure 20.7: The residential desirability surface for English-speaking students at Pointe Claire in Montreal. Although in Quebec, the surface is markedly 'warped' towards English-speaking Ontario.

their mental map is strongly skewed to Ontario, with peaks over the major cities like Toronto. What we have here is perfect geographic confirmation of the cultural and linguistic division that is of such great concern to all thinking Canadians on both sides of the 'border'.

The sense of Us and Them, the sense of the 'stranger', seems to permeate all human societies, and one wonders sometimes what can be done to create a world in which people are more open, tolerant and respectful of others around them. We saw in Chapter 14 the strategic paranoia of a Venezuelan army officer for tiny Guyana next door, with no navy, no air force and only a minuscule 'army', but almost every country appears to feel threatened by someone – both rightly (for some threats are distinct possibilities), and wrongly (for others are totally remote). A Kenyan army officer, for example, confines his strategic space entirely to East Africa (Figure 20.8), living in a world in which the United States, the Soviet Union, China and Europe do not exist in any significant way to him as a military man. Only Ethiopia and Sudan are friendly, and the greatest threat is seen as coming from Somalia. One wonders what sort of information is moulded into these varying images of threat and friendship, likes and dislikes.

It is not just army officers who transform information and ignorance into positive and negative preferences. We all do it in varying degrees, and sometimes the varying perceptions we have of geographic space are translated into actual decisions, such as folding our tents 'here' and moving 'there', otherwise known as migration. In the United States, for example, groups of people at various locations (in this particular case Pennsylvania), generate highly uneven information surfaces, with the

246

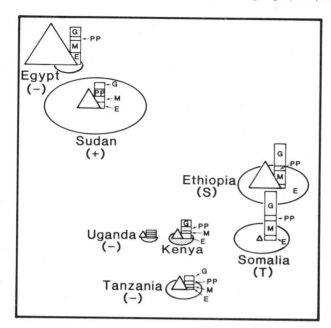

Figure 20.8: The strategic image of a Kenyan army officer. Major powers do not 'exist' in any military sense, but a threat is seen from neighbouring Somalia.

highest peak around their local area that they know best, falling away to the valleys that are areas of considerable ignorance (Figure 20.9). These hills and valleys of information are highly predictable from the simple gravity model ideas we have met throughout this book, for information tends to decline with distance, yet increase with population. After all, it is usually people who create consistent and reinforcing streams of information, even though environmental events like floods and earthquakes bring places 'into the news' from time to time. These events, however, tend to be random 'noise' on the stronger, and much more consistent information 'signals'.

Now the fact that the information surfaces are highly predictable should make us pause, because in a rather deep sense it says we are all prisoners of our location. To a very high degree, the information we have depends upon where we are. This should make us very thoughtful, not only about our location in geographic space, but also in many other 'spaces' – religious, cultural, racial, linguistic and so on. It should also make us think hard about the manner in which we translate such information into our quite personal preferences. For example, young adults in Pennsylvania generate a mental map or preference surface in which the highest peak is over Colorado (they were university students), although Pennsylvania is the next highest peak on the map (Figure 20.10). Generally, there is a

247

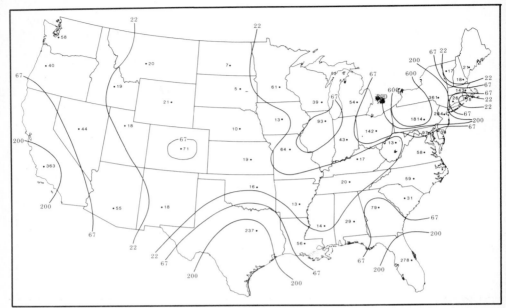

Figure 20.9: The 'information surface' of university students in Pennsylvania. A high peak (1,814 units) lies over Pennsylvania itself, but this falls rapidly away with distance, rising again only where large numbers of people generate more information.

bright image of the west coast and northwest, but it is downhill all the way to the Utah Basin and the low midwestern Trough. Texas is another peak, with steep gradients down to Mississippi and Alabama, until the surface rises again to the Carolinas and the northeast. The correspondence of the mental map with the information surface is not perfect, but most of the highs and lows fit fairly well. Some states – Colorado, Oregon, Vermont, Maine, Washington and New Hampshire – are more highly preferred than their information would lead us to expect, but these are all states with images of mountains and lakes and forests and wilderness . . . highly attractive images to an environmentally aware generation. Others – Ohio, New Jersey, Illinois and Delaware – are evaluated much lower than we might expect from the information, but these are perceived as old industrial states of dirty concrete and polluted air. And notice, whether they are or are not that way in reality (again, what do we mean by reality?), they are *perceived* as such by a generation of young, well-educated, and highly mobile Pennsylvanians. Many of these are going to be 'voting with their feet', and you can see now why the winds of influence (Chapters 13 and 17) blow out of the states of the old manufacturing belt towards those with newer and brighter images.

248

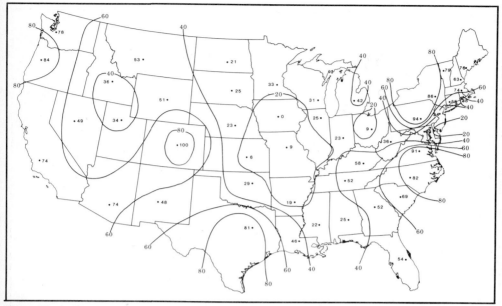

Figure 20.10: The preference surface for university students in Pennsylvania. The highest peak is over Colorado; the lowest over Iowa.

Yet even these images of residential preference are not wholly stable, and over the past twenty years some intriguing shifts have come about. If we look at the differences between two mental maps of the 1960s and the 1980s (Figure 20.11), negative scores blanket nearly all the northeast and midwest, *and* both Florida and California. The latter are only moderately regarded – 'nice places to visit, but I wouldn't want to live there' – because their once-golden images are now somewhat tarnished. The greatest increases have taken place in the Carolinas and Georgia – the hi-tech image of the Research Triangle and Swingin' Atlanta – and notice that this is just where William Warntz's steadily rolling income front (Chapter 13) would lead us to expect considerable change.

Although much of the early research in mental maps was carried out by geographers twenty years ago, it is only now that these ideas are trickling out into a wider world. People are slowly beginning to realize that it is the mental images that people hold of places that may well be more important for making decisions than any amount of 'scientific' census data. In certain important areas, such as tourism, the mental images are all-important, and colourful posters, brochures, and 'tourist literature' of all sorts are produced to enhance the images of

249

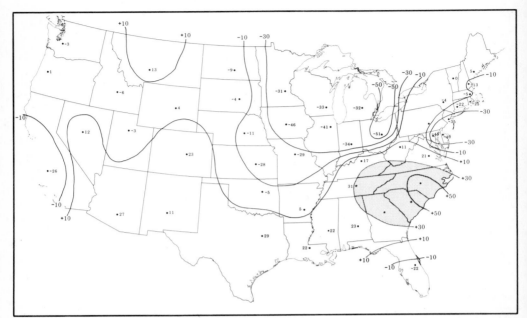

Figure 20.11: Changes in the preference surface of Pennsylvania university students between the Sixties and the Eighties. Negative scores blanket the midwest and northeast, while the Carolinas and Georgia have greatly enhanced their images.

places. Enormous flows of money are generated by the tourist trade, and many regions depend heavily upon the seasonal influx of vacationers to keep them going. In this rather hard economic context, the question 'what do people think of us' becomes more than just a casual inquiry. In France, for example, the regional images of a thousand people all over the country blend into a rather consistent image of likes and dislikes (Figure 20.12), and we can easily see why many of the *grandes routes* of Napoleon focused on Paris are crammed on 1 August. For many people, summer vacation means an explosive centrifugal 'push' to the periphery, to Brittany (Bretagne), the Pyrenees, the Alps, and the countryside and beaches of Provence-Côte d'Azur. And in between? Well . . . nothing much, because 70–80 per cent of the French people admit they do not really know what is there. Driving hell-for-leather to the high spots on their mental maps, perhaps at night 'to avoid the traffic', they keep their eyes glued to the road looking neither left nor right. Yet, as the poet Hilaire Belloc attested in his *Road to Rome*, a beautiful account of a walk through France, some of the loveliest country is in the Vosges and Jura mountains, the region that in this jet and automobile age

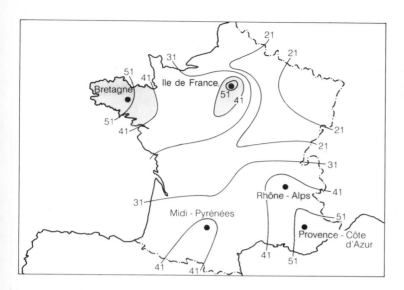

Figure 20.12: The 'knowledge' surface of the French people that translates directly into vacation preferences and the prices of new residences in France.

receives the *lowest* score of all. With all this 'mental image pressure', it is hardly surprising that the French surface matches almost perfectly another map of the price per square metre of new housing in France. Mental images are often translated directly into monetary terms.

How do you enhance the image of your region to attract more free-spending visitors? And, perhaps more importantly, do you *want* to attract great hordes, littering your beautiful countryside with all the plastic, glass, aluminium and paper refuse of an insensitive urban 'civilization'? Yet few can resist the attempt, and tourist ministries, offices, bureaux and organizations pour out a steady stream of image-enhancing material. In the United States, a very detailed and careful analysis was made of the images contained in the standard 'tourist packages' issued by each state. Some states try to promote a wide variety of images – the 'we've got something for everyone' tactic – while others focus attention on a few things that make them 'different' from the rest. In this sense, Alaska, California, Florida, Hawaii, New Mexico and New York appear somewhat more eccentric than the other states. The trouble is that when 50 states are pushing themselves, it is very difficult to present yourself as uniquely attractive, although some clearly do attract more tourists than their fair share, which immediately begs the question of what we mean by 'fair share'. Once again, this lands us right back in the gravity model, for we are talking about spatial interaction as people move for vacations to other areas. Distance, perhaps translated into costs and time, is obvi-

251

ously going to enter the problem, and we can construct a rough index of attractive success out of the interaction potential of a state and the actual numbers of tourists visiting it (Figure 20.13). Maine, Connecticut, Maryland, Virginia and West Virginia in the east, Mississippi and Alabama in the south, and Montana, Wyoming and the Dakotas in the northwest are clearly not achieving their potential. But perhaps they do not want any more people? Perhaps there are quite enough summer visitors to Maine and the Tidewaters of Maryland and Virginia? Perhaps the vacation environment is already saturated and 'overloaded'? These troughs of low success contrast prominently with the peaks of California, Florida and Texas, who are clearly 'overachievers' in the tourist stakes.

Figure 20.13: The relative success of states in attracting tourists. Some seem to be able to attract more than their 'fair share', while others do not reach their tourist potential.

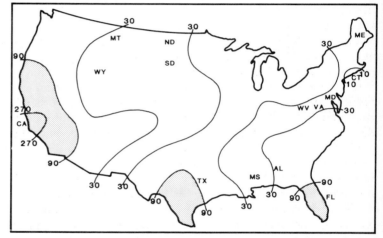

The question of tourist overload is not a trivial one, because wilderness areas are becoming increasingly vulnerable to the hordes of the uncaring, the weekend visitors and annual vacationers who so seldom feel they are responsible for the mess they leave behind for the 'locals' to clear up. Once again, it is that question of respecting and caring for others living in places far from our everyday lives. The few pockets of tranquillity and peace we have for genuine re-creation can be preserved only if we develop a sense of gentle caring for a world that is the heritage of all of us.

Children as geographers

21

When an old field like geography suddenly rejuvenates itself, and produces the sort of intellectual explosion we have seen over the past thirty years, the new ideas slowly but surely diffuse outwards and begin to influence the way others see the world around them. As a matter of fact, I would like to think that this book you are reading now will help you see geography and your world a little differently. Some children, fortunate enough to live in countries where geography is properly taught in the schools, have been strongly influenced by the changed outlook, and the new ideas have trickled down from research programmes and universities to the schools. Since many deep ideas tend to be fundamentally rather simple (perhaps really good ideas often are), they have been incorporated into the ways we teach children about the spatial and geography dimensions of their lives. Obviously the ways in which the ideas are presented have to be considerably modified, and their mathematical expressions reduced to simple counting, but the essence of an idea can often be preserved, and we must remember that even quite sophisticated forms of mathematics can be reduced to step-by-step arithmetic. Above all, we must never underestimate the children, or be condescending to them: once they are involved and intrigued with something, they have a capacity to 'catch on' with a speed that perhaps you and I wish we could somehow recapture.

Of course, children start learning about the 'geography' of their world long before school starts, and perhaps some of the most important experiences take place before formal schooling ever begins. The French psychologist Piaget has noted how babies start to relate things in the immediate space around them, and, once again, *relating* always means connecting things up to form structures – structures we continue to create all our lives to 'make sense' of the world we find ourselves in. Later, as a baby begins to crawl, there is a sense of deliberate exploration, of moving out of the immediate confining shell created by muscular weakness and immobility, to see what is on the other side of the hill. That 'hill' may only be the sofa or armchair in the living room, but as adults we must try to recover the sense of scale that a small child experiences. In a park in Stockholm, there is a bus built at 'child scale' for grown-ups, a bus so big that adults have to climb up and scramble on board, and then climb again to perch on the high seats on which they can stand up to peek over the backs to see the people behind. Do you remember how big those buses used to be?

It is important to remember such feelings if we are to understand a child's world, and use our understanding to open up a larger world to children. There are often caves behind sofas where you can be a bear, and later on rumpled carpets make wonderful hills and valleys for the elaborate disposition of soldiers – not to say secret tunnels through which you can sneak up on an unsuspecting enemy. Denis Wood spent hundreds of hours watching children play in dirt and sand piles in summer, and in the snow drifts of winter, and he recorded the elaborate processes of imitative learning and imaginative construction going on. 'Little kids' (less than 5 years old), watch and learn from the 'big kids', who are usually rather tickled by having a small and attentive pupil, and often take their teaching responsibilities rather seriously. At first it is just trucks and bulldozers, with roads and railway crossings (splendid places for crashes!), but then more elaborate landscapes evolve with carefully marked entrances and exits, long tunnels, holes and hills, and rivers and lakes. An unspoken, but acknowledged hierarchy of themes emerges: perhaps a general setting of 'trucks', within which various forms of play are considered proper by the children, while 'cake and pie making' (another distinctive form) are not. And the nice thing about dirt, as a 5-year-old girl solemnly assured Wood, is that 'all you have to do is look under the ground and there it is!'

Denis Wood
North Carolina State University
1945–

But whether sofas or chairs, doll houses and kitchens, and sand piles or snow forts, in all cases young children are creating a world at their own scale, or at a scale in which *they* are the big people, the ones who look down and see things from above. So perhaps we should not be too surprised to discover that even quite young children have a remarkable capacity to understand maps and air photos – 'it's just as though I'm a bird flying over and looking down,' they say, taking everything in their stride. And why not? *They* have been creating 'flying over and looking down' worlds for a long time – where 'long time' may mean two years of playing in the dirt pile, and two years may be nearly half of their lives. What is sad is that the ability to read maps, so seemingly 'natural' in young children, often seems to be partially lost once they get to school. Well-meaning 'pedagogues' make complicated what was obvious to any child, perhaps because the pedagogues (I like to keep the word 'teacher' for something better) have forgotten how to learn from children.

As children get older the spatial bounds widen, even as the sense of time becomes more acute. We are all geographers and historians, and perhaps the true teacher captures the sense of excitement in widening these fundamental dimensions of our existence, while the professional pedagogue crushes everything down, reducing the marvels of our world to checklists of capes and bays and dull dates, to be memorized for the next quiz or examination. One of the wonderful things about terminating your formal education, whether at 16 from school, or with a PhD from a university, is that no damned and boring old fool is ever going to examine you again. At this point, if it has not been all crushed out of you, you can start learning once again like a child – for the sheer marvel and pleasure of it.

We do not know very much about these expanding geographies of children, partly because we seldom think it worthwhile recording what it was like when we were children, and partly because our formal geographic research usually leaves the child's world behind. Only today are we beginning to get some glimpses of the sorts of things that shaped geographers in childhood, in part through a novel biogeographic project led by Anne Buttimer and Torsten Hägerstrand at the University of Lund, and the latter's gentle and exemplary 'reaching back' to recollect and reflect upon his own childhood. As for research that displays a real care and concern for children, I can think of only a few studies that demonstrate such qualities, nearly all by William Bunge and the students of his remarkable Detroit and Toronto 'Geographical Expeditions'. One map shows the density of child-maiming debris in Toronto (Figure 21.1), the broken glass and jagged tin cans in the vacant lots where poor kids play. Others focus on children in Detroit: one on rat-bitten children in the slums (Figure 21.2); another on the places where commuters have killed children with their automobiles on unprotected streets (Figure 21.3); a third dramatically recording infant deaths in terms of rates characteristic of poor Third World countries, rather than a rich America (Figure 21.4). Too seldom is the geography of the child's world the subject of such searing portrayal.

In nearly all societies, the time comes when children go to

William Bunge and
Joselyn
*Société pour l'Exploration
Humaine*
1928–
1972–

255

Figure 21.1: The density of child-maiming debris in a part of Toronto – the rubbish of tin cans, nails, broken glass, etc. that can cut and hurt a child playing.

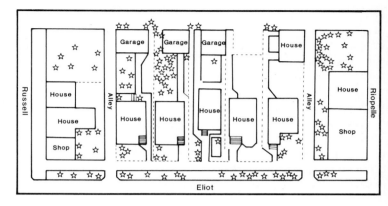

Figure 21.2: Where children have been bitten by rats in Detroit. It is estimated that many such incidents are never reported to appear in hospital records.

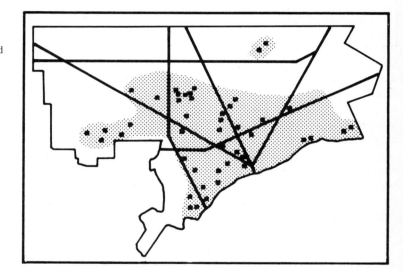

Figure 21.3: Where children are killed by automobiles in downtown Detroit. Each day commuters stream in from the suburbs and children playing in unprotected streets are often run over.

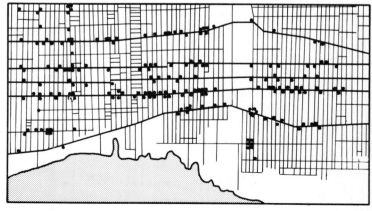

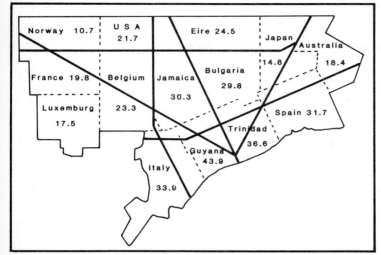

Figure 21.4: Infant death rates in Detroit recorded in terms of similar rates in other countries. In the centre (predominantly black), the rates are comparable to the highest in poor Third World countries; in the suburbs (predominantly white) the rates are among the lowest in the world.

school, and here they meet more formal and prestructured ways of learning. In some countries, notably the United Kingdom, the teaching of geography in the schools continues to be influenced by the (r)evolution that we have been looking at in this book. In other countries, it is still stuck at top dead centre, exactly where it was thirty years ago, or it has disappeared with history into the hodge-podge labelled 'social studies', an amalgam that leaves most children ignorant of both their historical heritage and the contemporary geographic world of which they are a part. In France, the appalling lack of historical knowledge has aroused even the Prime Minister, and history is being restored as a primary subject in the schools. Geography has yet to reappear, although years ago Jean Tricart warned the pedagogues about the intellectually numbing consequences of checklists. In Spain there is a renewed concern to revise the old curriculum radically, a geographic 'house-cleaning' long overdue. In the United States, highly consistent maps of ignorance surfaces (Figure 21.5) can be constructed year after year from *university* students, demonstrating that half of them cannot even identify some of the states in their own country, let alone other nations – many of which are regularly in the news. What does it mean in a democratic society when a newscaster reports 'The Senator from Wyoming [or Missouri or Arkansas] proposed an amendment in Congress today', and half the university-educated population do not know where Wyoming or Missouri or Arkansas are? Not long ago, many children used to play with puzzles in which the pieces were the states (or counties, or *départements*, or *län*),

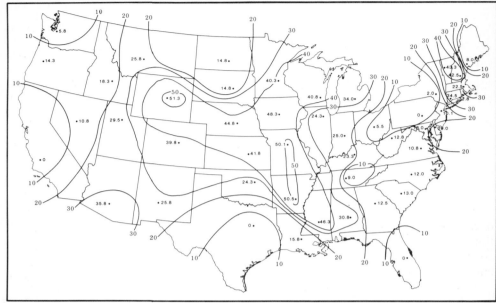

Figure 21.5: An ignorance surface of Pennsylvania university students in the United States. Fortunately, everyone knew they were in Pennsylvania, although more than half (50% iso-ignore) could not identify Wyoming, Missouri and Arkansas.

John Peter Cole
University of Nottingham
1928–

John Beynon
Trent Polytechnic
1945–

but today television watching has virtually destroyed such play activities from which so much was learnt on the way.

Fortunately, the picture is not dark everywhere: here and there we can see the dedicated efforts of some geographers to reform curricula, to generate opportunities for teachers to retrain, and to provide children with materials to explore some of the rich concepts and principles that help them to make sense out of the world. One of the most remarkable efforts is a series of four slim booklets called *New Ways in Geography* by John Cole and John Beynon of Nottingham. Written and designed for young children, perhaps 7 to 10 years old, they start by translating bird's eye pictures into maps; for example, a fishing village on a lagoon in Benin (Figure 21.6), or perhaps a small town in England that they are more familiar with (Figure 21.7). Children are asked to find their way on these maps, and their journeys gradually widen to the regional, national and continental scales. This seems a natural and sensible way to teach, to start with 'small' or local places, and then work outwards towards a larger world. One French geographer, Guy Thouvenot, has shown how people seem to classify places farther and farther away in terms of familiarity and frequency of visits (Figure 21.8), yet another example of that distance

258

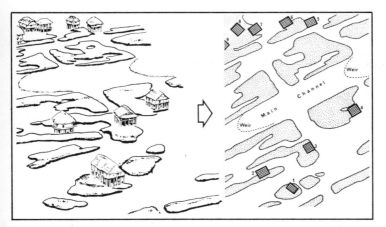

Figure 21.6: A bird's eye view of a fishing village in Benin, and its cartographic representation used by young children to answer questions about how they would find their way in this watery world.

decay effect, and his scheme matches the 'coquilles', the space-time shells of the psychologist Abram Moles, who was concerned with the interrelationships between daily rhythms and spatial scales.

Naturally, the children cannot travel very far in the wider world (although most teachers dream of taking their children on field trips overseas), but sometimes information about the world comes to the children in other ways. The children are invited to make collections of places from the labels on food containers, or from postmarks on letters saved by relations and neighbours. These contributions are pooled in class, and carefully located on maps. Locations, of course, have to be carefully designated, so the next booklet starts 8-year-olds on

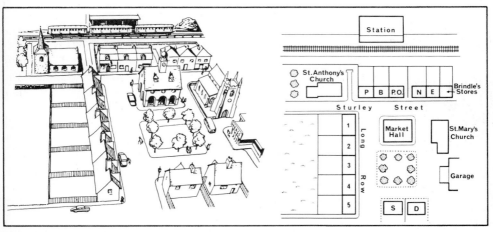

Figure 21.7: A bird's eye view of a more familiar scene to British children. Identifications have to be made between this view and the more abstract map representation.

259

Figure 21.8: The distance decay effect appears in terms of people's familiarity and frequency of visits to places, and this matches the imaginative space-time 'shells' we seem to live in. These constructs suggest that children might start to learn about a familiar world 'close at hand', and gradually widen their experiences to more distant places.

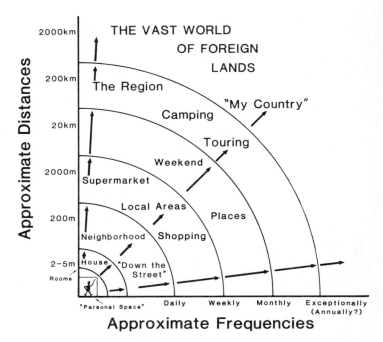

coordinate systems, starting with the rows and columns of desks in the schoolroom (Figure 21.9), but actually teaching the children what mathematicians call 'matrix notation'. More journeys are taken, this time over road, rail and air networks, sometimes

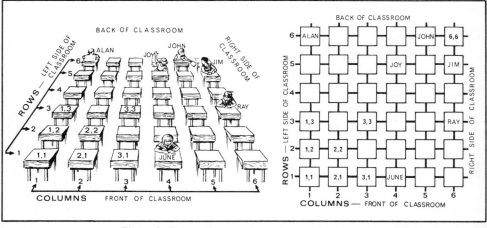

Figure 21.9: Locations can be designated by coordinate systems, i.e. so many units in one direction, and so many in another. Counting desks, and locating friends, is a good way to start, and translates easily into more complex ideas of latitude and longitude.

260

by quite roundabout ways caused by barriers (the Caucasus Mountains) or gaps (the Suez and Panama Canals). Eventually, the children have to build railroads over terrain consisting of marshes, forests, hills and mountains, but each square of land use on the chessboard map uses up different amounts of the children's resources (small counters). Blank squares on the map use up one counter, marshes two, and mountains four. The aim is to build the least-cost route, the one using the fewest chips. They are also asked to translate real road networks into geometric representations, and find their way over networks with breaks and blockages (Figure 21.10), perhaps to bring

County Down
Northern Ireland

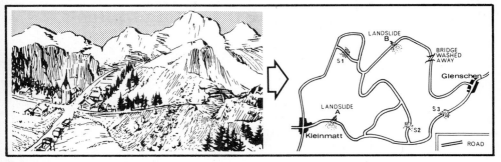

Figure 21.10: Route or 'maze' finding over networks that have been cut by natural hazards. How do we get food to Glenschen?

food to a small alpine village cut off by floods and landslides. Fourteen years later, in the university, these embryonic geographers may realize that they had their first lessons in graph theory, and in optimal and heuristic programming, when they were 8 years old!

The third book, now they are 9, starts them thinking about locations, and what they might mean by the *best* one. This sounds very familiar, because it touches upon that fundamental question of accessibility that we have looked at so often in this book. The children even start with the problem of selling cold drinks to people scattered along a beach, a problem taken straight from a classical mathematical analysis of spatial equilibrium in a linear market. They tackle the problem in simple, intuitive ways, but their conclusions are often the same as the mathematicians'. At this point, they go on to sets of things, and see how sets can overlap to produce a sense of connection as some things are related to other things. Recording vines and vegetables on flat or hilly land in a small two-by-two table (Figure 21.11), helps them to see how human choices may well be related to terrain and soil conditions, but no one tells them at this point that they are carrying out something that a statis-

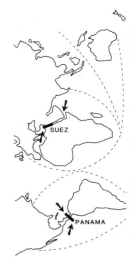

261

Figure 21.11:
Recording vines and
vegetables in a 2 by 2
table depending upon
whether they are grown
on flat or hilly land. In
this way children can
examine the degree of
association between
sets of things.

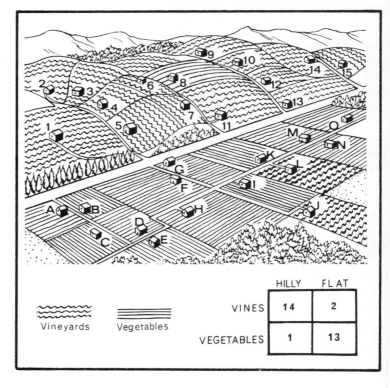

	HILLY	FLAT
VINES	14	2
VEGETABLES	1	13

Vineyards   Vegetables

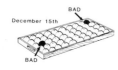

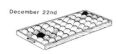

tician calls contingency table analysis. That is much too difficult for children!

From these simple countings of overlaps, they move to graphs – travel times versus distance, rainfall versus height in a mountainous area, slopes and potato yields in Peru, and how the intensity of land use may vary in a quite regular way with distance from a farmstead. Later, in the university, they will learn to call it regression analysis, and realize that land use is often influenced by distance, as we saw in our model of a city in Chapter 3. Even processes of spatial diffusion are introduced, starting with boxes of apples spreading rot from one to the other, but continuing with exercises that let the children diffuse seeds of trees, colonize islands, and eventually play the roles of a farmer and a nature conservationist. The farmer tries to diffuse his ownership over an area, and reclaim as many acres of land as possible for farming; the conservationist tries to diffuse public ownership, and protect as many acres as possible for a wildlife habitat. It is a grim game, played in the real world of Britain every day, and perhaps it is wise to teach the values underpinning such a 'game' early on.

By the age of 9 or 10, the children are formally introduced to map scales and topographic sheets, the idea of optimal paths through networks, quite sophisticated problems of optimal location, and some principles of classification. Wherever possible, exercises are designed to allow personal experiences, current happenings, and readily available sources of information to be used. Children calculate their own travel indexes, as well as those of cooperative parents; newspapers are sampled to see how many times places appear in the news – all carefully pooled and mapped before being used in classroom discussion. Telephone directories are scanned to see how certain names have quite distinctive distributions. The Beynons, for example, have their stronghold in southwestern Wales (Figure 21.12),

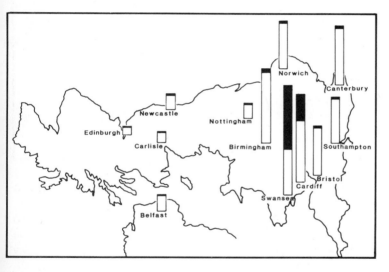

Figure 21.12: Distributions of names compiled by children from telephone directories. The Beynons seem to have their stronghold in southwestern Wales, while the Coles are scattered all over Britain.

and fade off in a strong distance decay curve around Britain. In contrast, the Coles seem to be everywhere. Finally, the children work with exercises on Brazil, England, Switzerland, Japan and the United States, the latter a game in which they try to work out the shortest way of visiting a selection of towns – what mathematicians call the famous travelling salesman problem, another of those combinatorial explosions we looked at before.

What these small booklets of exercises and games show is that children are perfectly capable of handling quite difficult and sophisticated ideas if only they are challenged in the right way. Exercises based on games may have an 'unacademic' sound to them, but if they are carefully designed the people who take part in them can learn a great deal. The gains are not just in factual information, picked up 'on the way' without rote learning, but also in a heightened intuition and a genuine feeling for situations and problems that is difficult to generate in any other way. So games can have very serious purposes, even if they are fun, and later in life some games are played very seriously indeed. Perhaps we should look at some much more complex and serious games that people (adults) play.

Playing games seriously

22

During the late 1960s and early 1970s, it became quite fashionable in many of the human sciences to devise games in which the participants took on various roles and learnt 'by doing', by interacting with other players, and by taking decisions which altered the conditions of the game as it went along. In geography, a number of such games were developed; for example, people had to build roads to search an unexplored map for agricultural and mineral potential. They could either plan lots of cheap feeder roads, which might break up under heavy traffic if the areas of potential were tapped, or they could invest in more expensive, but much more reliable routes, which had less chance of hitting an area for potential development. These sorts of ideas were also incorporated into games for young children, as we saw in Chapter 21. Perhaps the most famous game was one devised for high school students, who built the town of Portsville by playing the roles of people with various political and economic powers to influence and decide on the town's development.

Today, the crest of the 'gaming wave' seems to have passed, perhaps because too many of the games were a bit like Monopoly – amusing, but not terribly realistic, and ultimately a bit boring and childish. Perhaps such feelings mean that a game was not really well-grounded on sound knowledge of a real situation, and that it provided too few chances for creative and imaginative decisions. A game can be overstructured, so that the players are simply victims of the throws of the dice; or understructured, so that 'anything goes', and the game ends in chaos. Yet when a game is devised on a solid and detailed base of knowledge of real human problems, and when the rules provide just enough structure to let the players generate a large number of realistic developments and situations, it can often draw people into it so that they become highly involved emotionally and forget that it is 'just a game'. It is then that the experiences from the game really come through and teach people valuable lessons, even if there are some distraught moments during the learning process.

Such highly emotional moments, during which people display such things as anger, anguish, dishonesty, cynicism, fortitude, sorrow and chagrin, are very difficult to convey in a written description like this. After all, you say, it's only a game, and I certainly can't see myself getting so involved. If you have never taken part in such a game, I can understand that you might feel this way, so the only thing I can do is ask you to take my word for it. I have witnessed a number of such games, and I can assure you that people *do* become very involved indeed. This is true whether the game is played with students in English or American universities, with professionals in the Institute of Bank Management in Bombay, agricultural agents of the Agricultural College of Ishurdi, India, or people in the management training programme of the World Bank in Washington.

The game we are going to look at is called the Green Revolution, used today in management training programmes around the world, but especially at the World Bank where it has been used frequently to sensitize people to the problems and difficulties faced by small farmers of the Third World. The game was devised, extended and refined by Graham Chapman (Figure 22.1), who based many of the operations of the game on his meticulous and detailed fieldwork with small farmers in Bihar, India. His research included deep analyses of the proverbs passed on from one generation to another, sayings which incorporated a great deal of information about the environment in which the farmers had to survive. In many respects, such indigenous, or *ethnoscientific* knowledge seems much more pertinent to the local conditions than a lot of the so-called 'scientific' knowledge imported from the outside by well-meaning, but 'out of touch', western agricultural experts and local agricultural agents. For example, the local *nakshatra* calendar has quite different divisions of the year from ours, but these turn out to be much more suitable for designating the important times of ploughing, seeding, harvesting, and the arrival of the absolutely critical monsoon. Farmers recognize up to 50 varieties of local padi rice, and they rotate them so that diseases are less likely to take hold – things they also control with powdered ashes of certain trees, smoke, and the prophylactic use of manuring to insure strong and healthy plants. This local way is in marked contrast to modern recommendations of a few hybrid varieties, which require large amounts of artificial fertilizer and pesticides – things imported

266

Figure 22.1: Graham Chapman, who constructed the Green Revolution game from his extensive fieldwork with Indian farmers in Bihar, with a typical farm family – one of several that might represent participants in a particular run of the game.

and manufactured outside of the village which many farmers cannot afford. Rice also has multiple uses, so that the quality of the straw may be as important as the grain. In fact it is the women who are the real experts on rice varieties – although all the agricultural extension work is directed, or perhaps we should say *mis*directed, at the men. Even the goals of the villagers are quite different from what someone on the outside might expect. In a highly marginal and uncertain environment, the farmers do not direct their energies at all on farming for surplus and export, but much more on basic maintenance and existence, of simply making it through to the next year with a bit more than they had last time. Their lives are intricate and complex webs of social and economic relationships (some of which Chapman has analysed in highly formal ways), and it makes absolutely no sense to isolate purely economic considerations from their larger context, something that many Western economists still have great difficulty in understanding.

It was precisely to increase and enhance such understanding that the Green Revolution game was devised, to help people responsible for agricultural development programmes at all

267

levels to look at a difficult and uncertain world as it was seen through the eyes of small Indian farmers, rather than through the abstract world of mathematical models, shaped by theory devised in modern post-industrial society. If we start with one farmer, the kernel of the game looks deceptively simple. The farmer (F) uses technology (T) to try to influence and control the environment (E) to produce his crops (C). Yet underneath such seeming simplicity there is already quite a lot of complexity that can quickly develop into all sorts of possibilities or scenarios. The farmer's technology can include either the traditional or the new high-yielding varieties of rice, the addition of irrigation water drawn from wells, and the use of modern pesticides and chemical fertilizers. The environment can produce seasonal rains or dry spells, or hit the farmer with diseases, depending upon which cards in the environmental 'decks' come up. At the beginning of a game, farmer-participants are dealt land, family, wells and money, and then have to make the best of a world into which they have been cast by Fate.

Of course, a game is not played by just one farmer, but up to 20 farmers at a time, and they are allowed to interact in whatever ways they like – barring physical violence, not totally inconceivable during some rather tense moments. In fact, even the simplest version of the game (Figure 22.2) includes a game manager (GM), who takes on basic banking and trading functions; a commodity market (M), in which the output of the farmers influences the price they get for their rice; and a labour market (L). The influences of these realistic additions are given by tables compiled by Chapman from actual fieldwork, and they bear little resemblance to the sorts of simple mathematical functions used by theoretical economists. Even more complex games (for example, Exaction) have been devised, in which the government enters, the economy is divided into two (village and town), and so on. But given these basic components, which provide a minimal structural setting, the game is absolutely open and free. In fact, many people playing it ask in the early rounds, Can we do this? Can we do that?, worried that their idea of sharecropping, of forming a cooperative or a trade union, or even stealing and pilfering from their landlord or neighbour, is somehow not allowed. But there are no rules, except the inexorable march of the seasons, and the probabilistic input to the game in the form of the well-shuffled environmental decks. Players know how much time they have for each season and round (games are all-day affairs, and take up to eight hours), but when the game manager sees the clouds gathering on the horizon, and the monsoon comes . . . well,

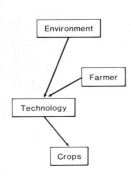

Environment

Farmer

Technology

Crops

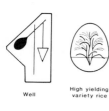

Well

High yielding variety rice

Fertilizer

Pesticide spray

268

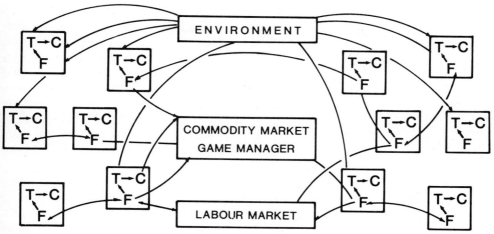

Figure 22.2: A schematic representation of a Green Revolution game with a dozen players. Not all the lines of possible interaction between the players are shown to avoid a too cluttered and complex diagram.

the monsoon comes, just as it does in real life. And if you have not prepared your fields in time (you were too busy organizing a cooperative, were you?) then tough luck – next year you and your family may well starve to death.

Out of this apparently simple game setting, an enormous number of situations can explode within the first couple of rounds, and some of these can be so distressingly realistic that a few participants have actually wept when they realized they had failed in the 'game of life', and lost their family one by one to starvation. For starvation is very real if you, as a farmer, do not have the resources of land and water to grow what you need, nor the money to buy food from others, nor enough surplus labour to hire out your children, and, on top of everything else, too many children arriving – yet another mouth to feed. If years are classified as good (A), reasonable (B) and bad (C), then a farmer who has been dealt a good hand by Fate may have sequences of As and Bs. But a poor farmer may only experience Bs and Cs, and it is critical to realize that a sequence such as CCBBBB is *not* the same as a BBCBCB. The order is all-important, for those two CCs at the beginning may mean that a small and marginal farmer may go under completely.

Not that the initial hand, the original resources at the beginning of the game, determine the outcome. Small farmers can grow into big ones, and large farmers can fritter away their resources. Excess land, meaning that you do not have enough family labour to work it, may mean you can hire labour (if you have the money, or are prepared to put some of your farm in hock with the bank for a loan), or you may find someone with excess labour who would like to sharecrop with you. Or you may join a cooperative, although these tend to take a great deal of time to organize, and often generate considerable disagreement as members argue over what the best policy would be this year. And always that awful sense of urgency is there: the seasons continue to march . . . the monsoon is coming . . . and, that's enough talk, we haven't got our fields ready yet! Others start organizing trade unions to bargain for higher wages with the big landowners in need of more labour for their fields. Unless there is a large surplus of labour (sometimes the case), these efforts usually succeed, but unexpected 'side effects' also appear. Unions may produce leaders who are not above feathering their own nests a bit (after all, they've got families too), and they seem to be welcomed by the big landowners because they represent an alternative and non-violent way of obtaining wealth. In a sense, the trade union leaders buy off an impending revolution generated by potential discontents whose energies are now channelled into union activities. Every once in a while revolutions are plotted in the game, particularly when Government enters but then is seen to do too little. At that point people start to think about taking Government over themselves.

When your family is going under, when your children start to die one by one of starvation, when you have failed to make it, then you too have little to lose. Many who are on the knife edge of desperation enter cooperative or union work, but at this point they may be bought off by the patronage of the big farmers. Failure to patronize poor neighbours ferments unrest, and many smaller farmers so patronized actually welcome their 'wages' in exchange for duties which allow them to steal or pilfer from their rich patrons. Although the game takes place in a room which allows free circulation, some farmers are clearly in the centre of things, while others are on the periphery, because even the micro-geography of the setting influences accessibility and player interaction. Information in the game begins to flow from farmer to farmer in highly structured pathways, and it is extraordinary how little each person really knows with everyone talking and bargaining at once. Oh, I didn't realize you'd lost a child last season, I was at the bank trying to get a loan. What a pity, we could have sharecropped together, why didn't you tell me you had extra labour? Accessibility and centre-periphery relations influence all sorts of interactions ranging from the clashing confrontations between

270

different classes of farmers who happen to be neighbours, to the virtual disconnection of a rich farmer sitting quietly on the edge of the game making money hand over fist season after season.

Even the settings in Washington, DC (the West), or Ishurdi (the East), or perhaps even a mixture of the two (New Delhi?), seem to influence the personal relationships and the degrees of cooperation and animosity. Western players tend to be much more individualistic, forming institutions to regularize and 'legalize' behaviour, although these tend to be rather unstable politically. Cooperation with another farmer usually takes place only as a coalition against a third. In contrast, Eastern players show much more cooperative behaviour, and use informal, and often unspoken agreements to regulate behaviour that tends to be quite reliable and stable. Young players seem much more callous and indifferent to death than older people, who are less grasping and more compassionate. Women players of both worlds tend to fight harder for the immediate preservation of their families, while men tend to look more to long-term strategies.

In the meantime, and totally unperceived by the individual players, the bank is quietly making a large profit. For short periods there may be antagonism towards the bank, when it forecloses on a small farmer who defaults on a payment, but these occasions are forgotten by the end of the game, when many small farmers are truly grateful to the nice bank for bailing them out, or helping them over a bad patch – one of those CC sequences produced by the environment that are so difficult to handle without additional resources. Sometimes the bank is too busy, or spends too much time working out loan collateral and conditions, and then some farmers may become moneylenders themselves, growing quite rich on exorbitant rates of interest, while others go up to their necks in debt. Once Government enters more complex versions of the game, and becomes an internalized part of the village structure, the possibilities for corruption soar, and many are soon realized. After all, it is worth bribing the bank manager or agricultural agent because the pay-offs can be so large.

Undoubtedly, an important part of the learning experience comes in the very careful debriefing sessions, conducted by Chapman or the game manager after an exhausting all-day session. Exchanges can be quite vituperative, with those who 'lost' blaming bad luck and the weather just like any small farmer, while those who did well put it down to their own skill and perspicacity and astute judgment. Slowly it emerges that

271

being a small farmer in Bihar (and in many other parts of the Third World) is a miserably uncertain business, and participants begin to sense the real tensions between what is happening to the level of output of the village as a whole, and what is happening to the individual farmers. No one comes out of the game – and remember that each one is different because of the freedom of the players to choose from so many alternatives – no one emerges untouched from playing it. One person said, 'These were the closest experiences I ever had – and I think will ever want to have – of being a small farmer.' I have seen students walking around rather quietly for a week after playing the game, and then suddenly exploding as everything they had been thinking about and bottling up inside them had to be talked about. I am sure they will never read about agricultural problems in the Third World again without making the Green Revolution a touchstone for the ideas, seeing if what they are reading 'rings true' in light of their gaming experience.

No single discipline can possibly illuminate, or even synthesize, all the complex elements going to make up a man-environment system such as a real farming community. But a geographer, with a natural and sensitive propensity for seeing connections between physical and social systems, can contribute to a deeper understanding of very difficult development questions. This is particularly true if the geographer's training and viewpoint are informed by modern areas of systems theory, and the methodological possibilities that these open up for describing complexity – often with the help of computers to handle the large amounts of information involved. It is for this reason that many geographers with these types of skills and training are sought after for consulting work, and, as I noted in Chapter 16, their contributions can be extremely valuable as independent sources of analysis and informed opinion. It is time we looked at modern geography and problems of Third World development a little more closely.

The geographer and Third World development

23

Your first impression may be that it is superfluous to single out problems of development in the Third World for special attention. After all, many of the examples we have already looked at come straight from academic research or consulting work in the Third World, ranging from agricultural and settlement schemes in Guyana, to the traditional environmental knowledge of farmers in India. These examples not only indicate the deep engagement of many geographers today in development questions, but they also point to what I can only call the 'utility conjunction' of much of modern geography. By that rather awkward term I mean that many of the themes developed in the recent (r)evolution of geography converge and join up here, and find deep application in the spatial laboratories of underdeveloped regions and countries. It is here that often real and pressing problems meet the new methodologies and research perspectives, which says something important about these ways of looking at old problems, and the work of those geographers whose skills allow them to undertake such practical and applied inquiry.

Many development questions have at their core fundamentally geographic answers – or perhaps I should say *partial* answers. The last thing I want to imply is that geographers can solve all the problems, for no one with the slightest sensitivity to the complexities of human-environmental systems in space and time would make such a claim. Simply to describe a large irrigation scheme, and raise the right questions about its under-performance, may take all the skills of geographers, engineers, soil physicists, meteorologists, anthropologists, and others in such fields as computer programming, systems and control theory, geology, agriculture, entymology, and possibly even economics. Nevertheless, many development problems do raise thoroughly geographic questions of location and accessibility, of central place networks and dynamics, of changing connections and linkages, of efficient flows of people and commodities, of the diffusion of ideas and innovations, of information obtained by remote sensing, and so on. Even questions of centres and peripheries arise, at a variety of scales, ranging from a villager's access to modern services, to the macro scale of centre-periphery relations between the First and Third Worlds. In brief, the exploding words we met in Chapter 3, *spatial*, *theory* and *model*, meet *regional*, *structure* and *planning* head on in many Third World settings of great practical application and utility.

Nevertheless, before we plunge into specific questions, I want to back off for a moment and consider, in very general terms, what happens when a country starts to go through the jolting and dislocating process we call, somewhat euphemistically, *development*. I say 'euphemistically' because the term *development* has a highly positive ring of nineteenth-century optimism to it (it is difficult to be against it), even though the consequences for individual lives can be horrendous as old traditions and relationships are broken, changed and perhaps crushed out of existence. Nor should we confuse mere growth, measured in a simplistic monetary way, with real development. A poor country can have a soaring income, perhaps because its huge iron or oil reserves are being rapidly exploited, but little in the way of genuine development that lifts the country as a whole, helps all the people, and provides a sustaining base after the main phase of exploitation is over. Everyone seems to live for today and forget about tomorrow, and this is as true of countries like Britain and Venezuela, who have foolishly squandered valuable oil resources in the short run for practically nothing in terms of long-term development, as it is of many poor countries. So the concept of development is more complex than it first appears, although it invariably means varying degrees of change and restructuring over geographic space.

When things change rapidly life becomes less predictable, which is another way of saying that things appear more chaotic and disordered. Suppose we think about a traditional and fairly isolated group of people: the rhythm of life goes on year after year, few new ideas are available to compete with the old, people play their usual roles in a well-known and accepted web of traditional relationships, everything is pretty stable and predictable and there are really few surprises. Contrast this with another group whose isolation has been broken by new roads bringing new goods, new ideas and new opportunities: literacy grows, more information is available, new markets

274

change old land use patterns, conventional wisdom is considered less relevant, relations between old and young, and men and women, alter under the impact of new ideas, and a road built to bring 'modernization' to the people now becomes the exit for the young leaving the rural ways for the bright lights of the town. Nothing is certain and sure any more, and life is full of surprises.

Following an idea of Bernard Marchand, we might equate high levels of surprise with uncertainty and disorder, and high levels of 'unsurprise' with predictability and stability. Then we arrive at a view of the development process as one that transmits pulses of disorder into stable areas that have to go through a period of chaos before settling down to another stable state, this time at a richer (in all senses of the word) standard of living – or what is the point? Geographically, this is close to an idea of Torsten Hägerstrand, that perhaps we can create maps of stability and change (Figure 23.1), maps of probability

Bernard Marchand
University of Paris VIII
1934–

S	S	S	S	C	C	S	C	S	S	S	S
S	S	S	S	C	S	S Town	C	C	S	S	S
S	S	S Rural	S	C	C	S	S	C	S	S Rural	S
S	S	S	S	C	C	C	C	S	S	S	S
S	S	S	S	S	S	S	S	S	S	S	S
S	C	S	S	S	S	S Rural	S	S	S	S	C
S	C	C	C	C	S	S	S	C	C	C	C
C	C	S Town	S	C	S	S	S	C	C	S Town	S

Figure 23.1: Areas of stability (S) and change (C), containing various amounts of 'unsurprise' and 'surprise' respectively. Both the traditional rural areas and towns are relatively unsurprising – if we know something about them we can guess much of the rest. But the areas in between, where change is occurring, may be areas of considerable surprise – knowing one thing about them may give us little information about the other things happening.

surfaces that evaluate whether various characteristics of an area will be predictable. Here the three towns are evaluated as stable, or unsurprising, which means that if we measure some aspect about them – say public water supply, infant death rate, per capita income – we stand a pretty good chance of predicting many other aspects about them. Similarly, the rural areas are also evaluated as stable: again, if we measure one aspect – road density, levels of nutrition, school enrolment – we can often guess the rest quite accurately. But in between, perhaps where the town meets the country, the areas are experiencing a very strong flux of change, and no matter what we measure, we have little chance of predicting anything else. The surprise value, that is to say the value of any information we can get, is very high. Of course, in this little graphic example, it appears that development is a mixed diffusion process, jumping down the urban hierarchy, and seeping contagiously out into the surrounding countryside. It is actually a much more compli- cated process, and we must remember that if we want to talk about the development process being transmitted, we cannot have transmission unless there is some *structure* to carry and shape it. Many geographers, coming from a variety of different intellectual positions, would hold that it is these deeper social, political and economic structures that must be understood, and perhaps altered, to further a development process and all that it can mean in terms of enhanced human lives.

In any development process there are certain aspects that can be carried out efficiently and in an optimal way – or, alternatively, in an inefficient way. And here we should not hedge or be mealy-mouthed: inefficient ways mean wasteful ways, and in a country like Tanzania, whose annual develop- ment budget is less than the amount New York City pays to dispose of its garbage, wasteful ways mean fewer schools, fewer hospitals, fewer . . . everything. Poor Third World countries understand perfectly the phrase 'optimum allocation of scarce resources', and usually want all the help they can get to make each penny do the work of two.

The problem is that raising questions of efficiency and opti- mization often scares the life out of some people, who see insensitive technicians manipulating the delicate webs of traditional life, and imposing computer-derived solutions regardless of the human consequences. For all my warnings about modern geography as a child of its technological time (Chapter 4), I can see very little evidence for such insensitive and manipulative bludgeoning. On the contrary, every geographic application of optimization and control theory I can think of in the Third World has been carried out with great care and sensitivity, and with genuine and ethical concern. For example, given a limited budget (and we see immediately that *all* forms of optimization theory focus upon making decisions under *constraints*), how do we upgrade a network of small feeder roads in a developing agricultural area to *maximize* their impact (Figure 23.2), particularly when they may be part of a larger development project, and when we know that the priority sequences we recommend will affect in turn many other pieces of the 'system'. This is a problem tackled by Thomas

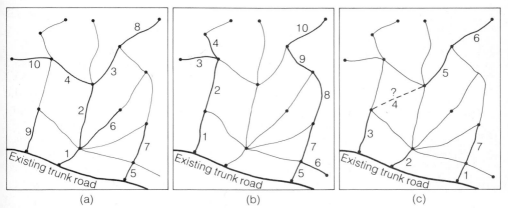

(a)   (b)   (c)

Figure 23.2: Alternative priority sequences (heavy lines) for upgrading existing rural roads (light lines) to higher standards, or building entirely new roads (dashed lines). The number of ways the priorities (1, 2, . . . 10) can be ordered results in a combinatorial explosion of possibilities from which the geographer has to choose the 'best'.

Leinbach in Malaysia and Indonesia, and is one of great complexity because it generates another of those combinatorial explosions we have met so often in this book. The alternatives (and I have shown only three out of the thousands of possibilities) all require an initial link from the existing trunk road, for there is probably little point in upgrading and increasing the capacity of roads in the region if they do not connect with a larger world. So we have to allocate the limited funds to certain roads and not others, and work out very carefully the sequence that will have the greatest impact. All this has to be done with an awareness that some links will cost more than others, so using up the budget more quickly, and if we choose to build a totally new connection (c), we may not have enough funds to upgrade as many of the other existing roads. Some areas will be more suitable for agricultural production than others, and we shall also have to be very sensitive to the fact that new or upgraded roads change the accessibility of the existing central places on the network. Like towns in nineteenth-century America competing for the railway, Third World villages want the new roads, and chiefs and headmen lobby politically to get them. Which is why an outside consultant can be so useful, providing an alternative viewpoint, and giving those who have to make the final decisions a useful scapegoat. Sorry I can't oblige you, but this is what the expert consultant recommended, and after all the Minister said . . . which places, of course, a heavy burden on the consultant to make the most careful analysis and to give the very best advice.

   Fortunately, it does not have to be the case any more that a

Thomas Leinbach
University of Kentucky
1941–

277

Gerard Rushton
University of Iowa
1938–

technically trained 'outsider' comes in and simply recommends a computer-derived solution for a region. Today, new optimization methods called goal programming mean that the geographer and local planners can discuss resources, needs and sequences, and try out various alternatives using formal optimization methods to simulate them. We would like to achieve these goals in this order, say the local people. Fine, says the geographer using goal programming, but the optimum way of achieving them with the limited resources available means you must go for *this* solution. Oh, good heavens, we don't want *that*, say the local planners. Very well, says the geographer, you will have to reassess your goals and priorities. So in this give and take way, highly formal and computer-derived solutions become a means of creating a dialogue. Various possibilities are worked back and forth until a satisfatory compromise solution is achieved.

In many development plans, it is those explosions of combinatorial possibilities that are so often the stumbling block, for they require efficient algorithms and modern computers to generate and evaluate all the alternatives. For example, in Karnataka, India, government planners had the resources to add 14 more primary health services to the existing system of 55 facilities, small units that provide basic medical care at the village level (Figure 23.3). The problem was that right from

Figure 23.3: The locations of 14 new Primary Health Units in Bellary District, Karnataka, India, found by the planners, who had already divided up the region in such a way that the most accessible solution could not be found.

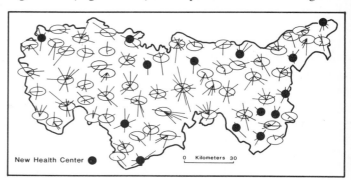

the beginning their approach virtually guaranteed an inefficient solution. With the best intentions in the world, they had divided up the region into population blocks of 15,000 people, each one of which would become a location for a primary health unit. But this original geographic division, which bore little relation to how people in the villages move and interact with each other, now became a constraint on the number of alternatives which could be examined. Gerard Rushton was able to demonstrate (Figure 23.4) that a computer-generated plan was

278

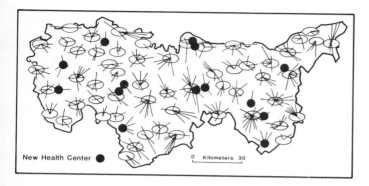

Figure 23.4: The solution found by a computer algorithm that was much more effective when judged by every criterion used by the planners themselves.

more efficient by *every* criterion the government planners themselves had used: average accessibility for everyone was increased by 11.2 per cent, the farthest distance a person had to travel was 8 kilometres less, and more villagers and more people were served.

The same concern for optimal solutions also arises on a much larger scale. One of the tragedies of our times is the Sahel, the desert fringe of West Africa, where changing weather patterns and human pressure on a fragile environment have produced increased desertification (Figure 23.5). In the early 1970s

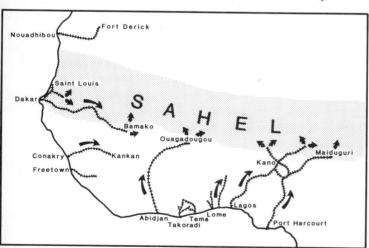

Figure 23.5: The Sahel of West Africa, where recurring drought has brought death and starvation despite major attempts to supply famine relief. Such problems linking origins (ports) and destinations (inland distribution centres) can be modelled effectively by computers to find the least cost solutions.

famine took tens of thousands of lives, and attempts were made to bring food to those who were starving to death. Although the intentions were wholly humane, the total relief effort is generally considered a fiasco. Those in charge had too little information about where the food was needed; about wharf, loading and handling capacities of the ports; about the road conditions (variable in wet and dry seasons); and what the

279

optimal pattern of movements should be. Devoted administrators literally broke down under the strain of trying to sort things out piecemeal, and normal customs procedures at the ports, based on the usual forms-in-quintuplicate, slowed departures to the point that one reporter noted that 'only the rats on the wharves of Dakar will be fat this year'. Food donated for starving children 'disappeared' into the private market sector, and bribery and corruption were rampant. Yet the overall geographic problem is a standard one in optimization theory: how to send commodities from points of surplus (the ports) to points of deficit (the inland distribution centres) most effectively when both port and road capacities are constrained. It should be perfectly possible to have computer models of such systems 'on line' as part of international contingency plans, and in the United States President Ford formally requested the National Academy of Sciences to evaluate the country's ability to respond to such large-scale emergencies. The usual sort of committee was formed, the usual sort of report was written, which said nothing that any newspaper reporter could not have said more forcefully, at which point it was promptly shelved and forgotten. When the next famine appeared (as it did in the 1980s), no one was better prepared than before. We have the ability to model large-scale geographic systems, involving complex interactions between the human and physical environments, and we certainly have the resources, which are trivial compared to the enormous waste generated by inefficient plans. As we bump from one 'unforeseen' crisis to another, all we lack is the will.

Any development plan, at whatever geographic scale we want to think about it, is concerned with connecting up and integrating a large number of different things, and it is hardly surprising that the 'systems approach' is coming more and more to the fore. One of the strongest advocates of such an integrative approach has been Akin Mabogunje, a geographer with a distinguished record of service in many areas of regional and national planning in Nigeria. In many countries it is a matter of mobilizing the people and the resources for genuine national development, not merely short-term growth, and helping them most effectively to work towards those goals the people want to achieve. But the question of goals raises in turn the values that a people hold, and these should always arise from the particular culture itself, rather than being imposed 'externally'. Whether the simplistic labels Capitalism, Socialism or Communism are used (and they are often so mixed up and ambiguous as to be meaningless), both investment planning and mobiliz-

Akin Mabogunje
University of Ibadan
1931–

280

ation are preconditions for any genuine *national* development. Whatever development goals are articulated, the chances of reaching them are greatly increased if those planning towards them are constantly aware of the connections between the parts, and how one part of a development plan can affect and influence another. 'Systems thinking' is almost mandatory today, and it need not imply at all that the people are being ignored. On the contrary, if a concern for creating more humane and decent lives were not centre stage, why bother to think about goals and development plans and projects at all?

Now if you take that last paragraph out of context, and then read the sentences slowly one by one, they could appear thoroughly platitudinous. Yes, yes, you say impatiently, integrated plans, connections, goals, planning . . . these, and other jargon terms, are all very well, and extolling the systems approach today is a bit like standing up for Motherhood and Country. Can't you be a bit more specific? How, concretely, might the systems approach shape an inquiry into a *real* and complicated system that might be of extreme importance to national goals? Fair enough: so let us go to India, and consider what is involved in controlling a very large human-environmental system like an irrigation scheme. And do not be put off by that word 'control', a word that seems to cause an almost Pavlovian reaction in many – although we would be wise to keep a careful 'thinking eye' on it. The plain fact is that somebody, somewhere, at some time has to make decisions about controlling an irrigation scheme, and the more informed and tighter the control the better. For one of the grave problems today is that the panacea of irrigation, extending both the area and the cropping seasons, is performing far below its possibilities, usually reaching only 20 to 50 per cent of its maximum capacity. Better control is essential for better performance, and better performance means better lives.

The first thing we see is that pieces of the system, and the people involved in making decisions about the pieces, are on different hierarchical levels. In a sense, the physical system is paralleled by a human system of control that tries to operate it. At the top are the huge dams and the high level engineers who control the flow of water, either for hydroelectricity or irrigation, two uses which may be incompatible since their demands may not coincide. Is it to be power for the cities today, or water to irrigate the farms tomorrow? Once released, the water flows through the main channels, some hundreds of miles long, and each of these may represent local subsystems. Eventually, the waters flow through distributary systems labelled branch canal, minor canal, field channel – each at a lower hierarchical level, until finally one of the 300,000 farmers opens his 'naka', the small outlet that controls (notice that we cannot get away from that word control!) the flow into a specific field. In a sense, the physical and human hierarchical systems of control are spread out across the landscape, but the connections and flows of information between the parts of the system may be highly tenuous. We may also know very little about lag times, seepage rates, defects in the system, soil moistures, crop requirements, the timing of applications, siltation rates, and the way salinity increases quickly with too liberal applications of water. Ignorance, that is to say uncertainty of control, increases rapidly downstream, and defective (or non-existent) communications and monitoring devices may mean a lack of feedback to those trying to manage the overall performance. Even overall goals may be inadequately defined, or be in conflict with one another, and no system can steer towards a goal if it does not know what it is. In India, it often appears that the parts of an irrigation system that are managed best are those where there are pleasant rest houses (accommodations for visiting officials) and good telephones – hardly surprising, because these areas then become the pieces of the system generating high information flows that are essential for effective management. It is worth noting that telephones installed in Sri Lankan villages immediately allowed the farmers to get 50 per cent more for their crops, and telephones provided to 146 Egyptian villages produced benefits 85 times the cost. Information is valuable.

We must also remember that systems physically decay – an irrigation scheme needs constant repair and dredging – and in a sense the parallel human system may also decay. Bribery and corruption are endemic in most poor countries (and are much more prevalent in rich countries than we care to admit). In large-scale irrigation schemes, contractors pay bribes to get the work of dredging and repairing, and since the value of land is greatly increased by irrigation it may be necessary to provide 'inducements' to those who control the water to favour one area over another. Large bribes are given by officials to obtain a particularly choice post as irrigation officer, and the size of these bribes is a measure of the monetary value of the posts themselves. It is usually considered insensitive and unpolitic ('you to not understand my culture') to discuss corruption in developing countries, but it constitutes a large and constant drain on very scarce resources that could be put to better use. In Ghana, for example, ex-ministers make amends after each military coup by repatriating their bank accounts from Zürich, for this sort of capital haemorrhage is common, and sometimes extremely heavy. Kickbacks of $30 million are not uncommon in Nigeria, and one businessman made a profit of $200 million in one year from contracts given as political patronage. One governor's daughter was able to save $5 million in her American bank account – a remarkable achievement of frugality. Monsieur Bokassa (formerly Emperor) is reliably reported to have stashed away $2 *billion*, a sum that represents many new roads, schools, hospitals, agricultural and industrial schemes. Any attempt to describe, and even model, a complex development project in systems terms will have to be aware of such seepage. It is not just irrigation ditches that leak.

Although we may instinctively think of agricultural problems in systems terms, this broad approach to structural description, control and decision-making is by no means confined to rural areas. Some of the most complex problems are to be found in the exploding cities of the Third World, and these cannot be effectively approached in a piecemeal fashion. With all the precious wisdom of hindsight, we can now see that the rapid growth of the cities, forced by high rates of natural increase and huge tides of rural-to-urban migration, is itself a partial response to investment decisions taken twenty to thirty years ago. Large proportions of all the loans provided by the World Bank went to basic 'infrastructure' – transportation and communication in many forms – to aid agricultural development in rural areas. The motives were impeccable, but practically no one in those days saw the implications for the overall 'system'. When you improve connections, you change the structure, and when you change the structure you should not be too surprised if the 'things' that live on the structure are redistributed. Migration to the cities has been so large that the World Bank, in a dramatic policy shift, has been virtually forced to give massive help in the area of urban planning – the provision of basic sanitary and sewage facilities, pure water, housing and health care. Each of these, of course, constitutes subsystems connected to others: for example, providing pure water to an exploding city means taking into account its geology, hydrology and soils, as well as the natural and human drainage systems. And drainage systems carry not just natural human waste (and many water-borne diseases), but increasingly chemical waste. If tap water in many American cities now contains traces of over a hundred chemicals not found thirty years ago, we at least ought to be able to avoid such contamination in the expanding cities of the Third World by proper planning and

283

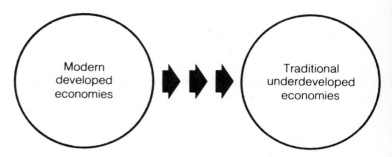

Figure 23.6: One view of the development process: a poor country is considered to be like a rich one, only at a lower stage, and eventually capital, technology, etc. will 'diffuse' to it.

Lakshman Yapa
The Pennsylvania State University
1940–

control as new and basic systems are laid out.

A consideration of something as basic as fresh water raises another important perspective on the development process. It can be strongly argued that many large-scale projects, even when successful, tend to benefit a small upper and middle class, and that theoretical 'trickle down' effects either remain theoretical, or perhaps work so slowly as to be almost imperceptible. It is for this reason that we can argue for a 'from the bottom up' rather than a 'from the top down' approach, focusing much of the development energy on what Lakshman Yapa has termed the 'basic goods economy'. This starts with the basic goal of providing food, shelter, clothing, health and basic functional literacy for all the people of a country, and recognizes that for all the huge development schemes of the past twenty years, two-thirds of the world's people have only made gains of less than $1 per year, and the poorest 40 per cent are now actually worse off than they were before.

The problem stems from the traditional views of the development process that too frequently detach all the purely economic aspects from the social, political and ecological. A poor, traditional and underdeveloped Third World country is considered to be like a modern and developed country, only at a lower stage (Figure 23.6). 'Development' is simply a matter of diffusing capital, innovations, technology, etc. from the First to the Third Worlds, a sort of development by imitation in which eventually people copy the patterns of consumption of the rich countries. The problem is that one, even quite modest basket of goods for a middle-class person is equal to 6–8 baskets of 'essential goods', or to put it more directly, one middle-class person represents an increase of 6 ordinary people. These habits of thinking that come with 'imitative development' carry much more widespread implications. Labour-intensive technologies and systems of production are downplayed, or ignored altogether, and technological fixes like the Green Revolution (Chapter 22) replace ecologically sounder agricultural methods.

284

As we have seen, special seeds, irrigation, 'hard' fertilizers and ecologically disturbing pesticides all favour the big landowners, and in fact the gap beween them and the poorer farmers has widened. Moreover, food crops grown to raise luxury foods like meat are extraordinarily inefficient: one hectare of land producing foodstuffs for chickens can provide the minimum daily protein requirements for 1,430 people. If it is used for soya, consumed by people directly, it can provide the same levels for 22,700 people. It is the same sort of story in energy generation and fertilizer production: in India, 21,150 small methane plants can be distributed widely over a region to produce 6 million megawatts of energy, 230,000 tons of nitrogen fertilizer, and employ 13,750 people for 1.07 billion rupees. A modern coal-based plant to produce the same amount of fertilizer *consumes* 0.1 million megawatts, employs only 1,000 people, and costs 1.2 billion rupees.

The fundamental error in most development thinking is the failure to see the Third World countries as pieces of a highly connected structure, rather than as closed systems into which development and modernization are diffused from some sort of

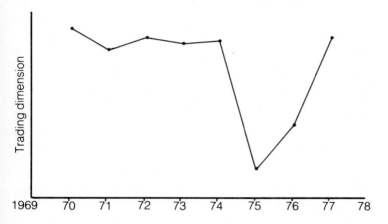

Figure 23.7: Portugal, as an international trading partner, was quickly 'crushed down', and had its connectivity and dimensionality reduced immediately after the 1974 revolution that ended Salazar's 40 year dictatorship. When it was seen to move 'left', but not too far to the 'left' it was allowed to expand and reconnect once again.

disconnected 'outside'. Every country is connected with every other, and sometimes the connections themselves can change drastically. Even a European country like Portugal can experience sudden disconnection. If we think of countries having different 'dimensions' (the United States is huge with many dimensions, while a country like Chad is very small with few dimensions), then the arrival of an important event like a political revolution may alter the dimensions drastically. After the 1974 revolution in Portugal, one of the least bloody and most peaceful revolutions on record ending forty years of dictatorship, the country was immediately crushed down by the rest of the international community (Figure 23.7). However, a year later, when it was seen that the new government was 'Left . . . but not *too* Left', Portugal was allowed to reconnect once again. International connections mean that development processes in the First and Third Worlds are intimately tied together, often in ways that reflect a colonial past (Figure 23.8). This can mean that every additional development in an already rich country can be at the expense of a poor country. Historically, the effects of British control over Indian agriculture were devastating for the production of food, as farmers turned more and more to export crops to feed British factories. Every cycle clockwise of Britain's development meant an anticlockwise cycle into deeper underdevelopment in terms of India's ability to provide basic goods for all her people. Similarly, in Sri Lanka,

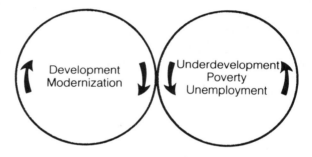

Figure 23.8: Another view of the development process: poor countries are tied to the rich world, and every turn of the development 'wheel' clockwise for the already-developed countries means an anticlockwise turn to deeper underdevelopment for the poor.

the British arrogantly appropriated 'unowned' (i.e. traditional and communally held) land to the Crown, and then sold it at bargain prices to British tea-planters. In the process, the farmers producing food were driven off, or virtually became tied workers to the system producing for export. Once again, we have a view of a dynamic system at work, but the question is how to change and control it to provide a basic minimum of goods for everyone.

There is one final aspect of the involvement of geographers with the development process that cannot be ignored – the role of university teaching and learning, a process and a potential that cuts both ways. For it is not just a question of people in the universities teaching those who come to them as students, but also learning from them, and assimilating the experiences of the Third World to create a deeper understanding of the problems and difficulties. It was after a seminar with agricultural officers from developing countries that Michael Chisholm wrote on relationships between rural settlements and land use, using a simple farming model as a lens to see a high degree of

Michael Chisholm
University of Cambridge
1931–

286

regularity in the apparent patchwork quilt of land uses characteristic of many tropical landscapes. Many who have taught
students from the Third World have had the experience of
learning as much as they could give, for it is here that theory
developed in one context often meets a rather different reality.

As always, the attitudes of those teaching, the values that
they demonstrate in their lectures, can be critical for those
learning. One of the major problems is that by the time many
Third World students reach the university they may have
absorbed some of the worst attitudes of intellectual snobbery
from their educational systems – systems often modelled on
those of the former colonial powers. Some of these are such
thoughtless and inappropriate transplantations that you wonder
how anyone in their right minds could have made them. At the
University of Ghana, for example, students used to run around
in worsted Oxbridge gowns, when it was 35°C in the shade,
and the Department of Classics taught Greek and Latin – but
no Arabic. Most of the students opted to do 'Arts', when the
country desperately needed engineers, soil scientists, hydrolog-
ists, agricultural biologists, poultry geneticists . . . and *geographers*, of course. It is difficult not to feel that the practical,
land-grant tradition of the United States would have been far
more appropriate.

Such attitudes of intellectual snobbery show up in distressing
ways. Learning is considered always to come from books,
rather than direct experience and observation, and the motto
often seems to be, 'Don't get your hands dirty.' Geography
students at one Third World university I know could hardly be
prised out of their library to do fieldwork, although an inspired
teacher at the University of Kumasi, on loan from UNESCO,
had already demonstrated how such students could contribute
to practical planning problems and their own understanding
simultaneously. In contrast, geographers at the University of
Zambia took a highly practical and involved stance from the
beginning, showing the students how a modern geographic
perspective could contribute in many ways to the development
of a country. Similarly, in Jamaica, many students work first-
hand with all the problems of urban geography engendered by
rapid rural to urban migration, for the capital Kingston contains
over half the entire population of the country today. They also
work on the social and economic aspects of land reclamation
and resettlement schemes after bauxite deposits have been
stripped away from a valley. The outlook for practical and
applied geographic work is much brighter today, but there is
still the need for geographers in the universities to think

through very carefully the sorts of curricula that can help Third World students (and their own) to contribute most to the problems of their countries.

Part VII

Thinking about what we think

Geography modulo the ideology

24

Thinking about how and why we think is not easy. It is perhaps the most difficult task we face as human beings, yet at the same time perhaps the task that ultimately distinguishes us as human. In all conscience, I cannot lead you on like a Judas goat, and tell you that the three chapters in this section of the book will be easy. I find thinking about the way geographers think a terribly difficult business, so these poor pages can only be a reflection of my own struggles. Some professional geographers may even say they should never have been written, that the ideas are too difficult, and still too nebulous, to present to a general audience as part of the recent story of geography. They may well be right. But the thinking by geographers about geographic thinking is such an important and vital part of modern geography that I feel a deep compulsion to try to lay out some of the main threads.

Paradoxically, the recent self-reflective concern is at one and the same time the most hesitant and the most confident. Hesitant, because only the truly foolish and ignorant are bold and sure in this old arena that we call the philosophic perspective, a perspective in which we see men and women struggling, over and over again, with the same difficult problems that have exercised their humane and imaginative concern for 2,500 years. But confident, too, because the thinking is underpinned by a deep sense of renewal of an old, important and deeply informing tradition. Human scientists, geographers among them, have not been a very reflective lot, and it is difficult to point to genuine philosophical concern, apart from a few token genuflections to the philosopher Immanuel Kant – who taught a very broad, almost all-embracing geography at Königsberg in the middle of the eighteenth century. However, Immanuel Kant deserves more than just token acknowledgments and a few cursory footnotes, for like all great philosophers he offers us, above all, an invitation to think. It is that invitation, renewed again and again as one of the deepest traditions of the Western world, that I believe has been taken up once again by a new generation of geographers.

Whenever a geographer asks questions and inquires into something – commuting patterns, irrigation schemes, urban renewal projects . . . any one of thousands of subjects that come under the scrutiny of modern geographic research – he or she always brings to the task a set of expectations and values that have the capacity to form and shape the inquiry. The expectations and values may have been thought about deeply, and the geographer undertakes the inquiry reasonably confident that the directions taken and the questions asked are the right ones. Or the values and expectations may be latent and unexamined, in a sense the taken-for-granted and 'naively obvious', which is perhaps another way of saying that they are not considered worth thinking about. But whether examined and accepted (and therefore unlikely to be thought about afresh), or simply accepted and unexamined (and therefore not thought about at all), the values and expectations form the broad frameworks of thinking within which each and every one of us inquires.

There is nothing very mysterious about this: *all* human beings are initially drawn into the culture, language, values and habits of thought of the particular world into which they have been thrown, and we all tend to 'see' the world, and inquire about it, through our own particular ideological lens. If we are human beings shaped by values it cannot be otherwise. The problem comes when our personal ideologies 'close down', and trap us into thinking that *our* way of seeing the world is *the* way of seeing the world – *the* way, in the sense of being the truthful, moral and *right* way of seeing the world. When this happens, it is only a short step to that strange and little understood human process of mythologizing our beliefs into something that closely resembles, and perhaps actually becomes, religious conviction. Religious conviction, by definition, is conviction based on faith, and such an act of faith inevitably means a closing down of real, and once open, thinking. Perhaps we reach the point when we feel we have finally found *the* way, *the* truth, and *the* light, and thank . . . well, Goodness, for that. No need to think any further: *the* truth is in front of us at last.

The result is that if we have geographers (or anyone else) carrying out research, and inquiring into a particular area, the results are always going to be shaped in varying degrees by the underlying, spoken or unspoken, ideology that the geographer brings to bear. Take a single, but highly complex subject (for example, an irrigation scheme), and half-a-dozen geographers with different ways of seeing the world, and you are going to get six different views and analyses, some of which will support each other's conclusions, some of which will complement the other approaches, and some which will appear to give you diametrically opposite results. It is a very nice question whether such inquiry into one aspect of the human condition can be labelled a *science* at all. Because surely a hallmark of any science is the creation of what we might call *intersubjectively shared and verifiable knowledge*. If we cannot agree, if we cannot share knowledge between us, and we cannot verify that knowledge in some way to our mutual satisfaction, do we have a science at all? Do we have to face the fact that the phrase 'human science' may be an oxymoron, a phrase that contains a contradiction within itself? Similarly, if six economists tell us to steer the economy in six different ways to six different goals (or even the same goal), are we totally confident that they know what they are talking about? On the other hand, is it conceivable that six physicists could give us six different views about the physical world?

It is not, of course, quite as easy as that. There have been numerous periods in physics and chemistry when disagreements were rife, and the results of the same experiment were interpreted in different ways. At the frontiers there is always disagreement, constructive if possible. The difference is that in our inquiries about the physical world of things, we often have the critical experiment available to us, and that eventually settles the issue – at least for the historically contingent moment. But in the human world questions are not so easily settled, we do not have quiet laboratories where we can hold some conditions constant and experiment, and anyway we should not treat people as things – and notice how my values are entering here. Creating intersubjectively shared and verifiable knowledge is much more difficult in the human world, where the expectations and values about that world inform and shape the approach to a very high degree. This is why the title of this chapter has that rather strange phrase 'modulo the ideology', and I think I owe you an explanation.

When we are children and first meet arithmetic in school, right at the very beginning, before all those awful fractions and decimals start to confuse us, we learn how to add with whole, or integer, numbers. Take any two integer numbers, add them together, and you get another one. The number you get is always available as an answer, because we use what the mathematician would call the set of all positive integers. This is an infinite set, so the answers we need are always available. But suppose you did arithmetic (again, just adding) on a finite set of numbers like {1, 2, 3, 4}; for example, 1 + 2 = 3. Fine, we have taken any two numbers, 1 and 2, added them together with that + sign, and the answer 3 is there. We try once more, 2 + 2 = 4, and again, no problem. This time 3 + 4 = . . . and, yes, we do indeed have a problem, because at first we think the answer is 7, and so it is if we are working with all those positive integers. But *that* answer 7 is not in our set, so that particular addition is not defined, and I think you will agree that it is a bit silly having a sign like + telling us to do something that works sometimes but not others.

Now the mathematician gets around this by defining her own sort of arithmetic, one that works all the time, not just sometimes. One of the arithmetics she can create with our set {1, 2, 3, 4} is called modulo 4 arithmetic, which says that when you use that +, then you must add the numbers in the usual way, but if the answer is not in the set, divide by 4 and take the remainder as the answer. For example, 1 + 2 = 3; 3 + 2 = 1; 3 + 4 = 3 . . . and so on. Notice that if we were to change the *base* of our modulo system, say to 3, we would get different answers: now 1 + 2 = 3; 2 + 2 = 1; and 4 + 4 = 2. The answers are always *modulo* the base of our system. So, in an analogous sense, are our 'answers' in geographic research, or any other area of human inquiry. Depending upon the ideological base we start with, we will get different answers – although we might hope for consistent answers (intersubjectively shared and verifiable knowledge) within any particular ideology.

There are many ideologies 'modulating' geography today, but perhaps the clearest example of the way a particular way of looking at the world shapes inquiry is the Marxist perspective. Equally, of course, a Marxist or radical geographer (the two terms are used almost interchangeably today), could say that the 'capitalist' geographer's perspective and values also strongly shape any inquiry, and this would be perfectly true. However, the Marxist would also probably claim that his or her way of looking at the world, and the set of informing values, had been examined and thought about very carefully in order to arrive at the true way of seeing, while everyone else went about their research and teaching rather naively without realizing how they were trapped. To the degree that all our ideologies remain 'latent' and unexamined, I think the Marxist has a distinct and valid point. Personally, I would have no problem whatsoever in marching forward shoulder to shoulder with my geographic comrades under the glorious banner of philosophic thinking, although when the Long March was over we might not have arrived at quite the same conclusions.

It is difficult to speak of *a* Marxist perspective today, because there is such a wide range of viewpoints which come under this label, but I think it is fair to say that most of them acknowledge the humane and ethical concern of Marx, a concern displayed as transparent outrage in many of his writings. In the middle of the nineteenth century, in the midst of all the humanly destructive turmoil of the Industrial Revolution, Marx looked around him as few others did, and was deeply and morally angered by what he saw. Steeped in the educated discourse of his day, particularly the writings of the philosopher Hegel, and influenced by the sense of logic and law to be found in the strongly emerging physical sciences of things, Marx also looked for social explanation in lawful historical processes, movements characterized by one period generating its own inherent and internal contradictions and changing into another under wrenching, revolutionary circumstances. Karl Marx was also a child of his time, just as we all are, and to point to this is not to denigrate Marx in any way. Only a fool would denigrate such powerful thinking in its own time and place. Thus, the underlying 'cause' of all the human misery was found in the capitalist mode of production, in the sets of economic relations between those who owned the means of production, and those who only had their labour to sell to the 'system'. This was seen essentially as a predator-prey relationship, crisis-ridden, and, some would hold, crisis-*dependent*, and the day would come when it would tear itself apart by its own internal contradictions and change into a higher form of human organization.

This framework, according to Marxist geographers, is the only true way of seeing, the only approach that lets us see the true reality. According to David Harvey, whose own work has illuminated many aspects of the urban crisis, it is 'the only method . . . [to] grapple with issues such as urbanization, economic development and the environment . . . founded in a properly constituted version of dialectical materialism as it operates within a structured totality in the sense that Marx conceived it'. Which is a very strong claim to the truth, and only a step away from a mythological transformation into such deeply held belief that it is at least reminiscent of religious conviction. Richard Peet, another strong advocate of a radical geographic science, would add that the Marxist framework leads to 'a radical program for restructuring society' which 'reflects the experience and the wishes of a reawakened people'. Which is

Richard Peet
Clark University
1940–

295

also an enthusiastic and exciting claim, containing not a small sense of messianic verve.

Now given the fact that all ideology shapes discourse, it is obvious that I cannot write about ideology shaping discourse without my own discourse – what you are reading now – being shaped by *my* ideology. Which should immediately put you on your guard to read carefully and critically. Moreover, if you agree with what I am writing, be extremely careful, and think whether you accept what I say simply because it supports your own preconceptions. On the other hand, if you disagree with what I write, again be very careful, and think whether you disagree simply because it does not support your preconceptions. I am not really bothered by agreement or disagreement, providing both make us think. Being lulled unthinkingly into agreement, or disagreeing unthinkingly, are each as bad as the other. That said, let us go forward together.

There is no question in my mind that the appearance of Marxist concern in geography, and its concomitant shaping of the lens through which the world is seen, has greatly enriched our methodological approach. There is an insistence that the things at the surface are not always what they seem, and that it is crucial to dig down underneath the superficial appearances to get at the 'deep structures'. I think this is quite right, and that it is important to have this stated again and again. It also happens to be a fundamental tenet of all good science, but one that tends to be forgotten too easily. There is equally no question in my mind that a sense of moral outrage permeates many areas of geographic inquiry that were ethically quiescent before. The notion that a geographer, or any other human scientist, is simply an ideologically cold, valueless, and 'objective' observer, standing aloof and apart, recording and analysing in a neutral and totally unconcerned way what is 'out there', was never true, although the mythologizing of science itself frequently gave such an impression.

Up to this point our discussion has been rather general and abstract, but it is important to think about some concrete examples where the Marxist concern has raised questions that would not have been considered otherwise. In the late 1950s and early 1960s as one country in the Third World after another was gaining its independence, there was a spate of 'modernization' studies – studies that tried to record how a newly emerging country had changed from about the turn of the century to 'modern times'. Processes of diffusion (Chapter 12) were often invoked, and both contagious and hierarchical notions were reformulated as 'growth poles' and 'trickle down effects'. Essentially, the 'modernization process' started at a few key places, and then 'trickled down' the urban hierarchy, eventually spreading out in rippling waves through the countryside. A Marxist would make the claim that such studies only looked at the surface appearances, sometimes in the quite literal sense of mapping the changing 'modernization surfaces', and they totally failed to raise even the most 'obvious' questions about relationships between the country and the colonial power. I personally think that much of such criticism is valid and unanswerable, and I say 'personally' because I was one who conducted such studies.

Richard Morrill
University of Washington
1934–

As for moral concern, a deep sense of justice permeates many areas of geographic research today, whether the practitioners would hold to the Marxist perspective in all its aspects or not. The research of Richard Morrill has consistently displayed a deep moral concern, and he was asked by the Supreme Court of the State of Washington to undertake the reapportionment of congressional districts, so that one person would truly represent one vote. Every politician knows that the way you divide geographic space can have a marked effect upon the representation of different political parties. In Britain, Peter Taylor and Ron Johnston have undertaken similar research, pointing out that although an Independent Commission may be neutral in intent, the particular districting plan it chooses will inevitably favour one party at the expense of other parties. Hence, it is unlikely that fair and just representation can ever be totally achieved by carving up geographic space. Ideally, the proportions of seats held by each party should reflect the proportions of people voting for them. However, since you never know (at least in a democracy) how people are going to vote, how can you divide up geographic space into districts *before* an election so that the results will reflect a fair distribution?

Peter Taylor
University of Newcastle-
upon-Tyne
1944–

As a part of his deep concern for the urban condition, David Harvey contributed to a study of the housing market and code enforcement in Baltimore, a city with large areas of decaying housing close to the centre, where 'redlining' was prevalent – the practice of private financial institutions outlining areas considered to be poor investment risks, where loans for building and improvement would not be given. Once an area is redlined, it becomes virtually impossible for a neighbourhood to improve its housing. Unknown to many real estate operators in the city who read the report, people who nodded their heads in sage agreement with the 'facts', many of the statements might have been written by Friedrich Engels looking at Manchester a

Ron Johnston
University of Sheffield
1941–

century earlier. Many geographers are 'involved' today, and their research, consulting and teaching reflect a moral concern almost unknown before.

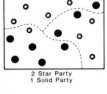

2 Star Party
1 Solid Party

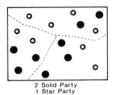

2 Solid Party
1 Star Party

The question of human research being shaped to a strong degree by the underlying values and ideologies also poses some grave difficulties for research programmes involving what we might call 'cross-cultural' perspectives. For example, suppose we want to know more about the enormous flows of television programmes that move between countries, flows that constitute a $1 billion dollar industry today, and carry huge loads of alien values from one culture to another. Swedish television, for example (Figure 24.1), draws widely from around the world for programmes of many different sorts, made in countries with very different sets of values from those characterizing much of Sweden today. The first thing that must be done in such research is to create very carefully defined sets of words to describe both the content of television programmes, and how that content or subject matter is treated – for example, for light entertainment, for informative or educational purposes, for children, and so on. Now suppose we have a set of programmes to be coded, and two coders: one a young American executive from a large national television network, the other a young Latin American academic whose views are deeply influenced both by the Marxist perspective and by the overwhelming dominance of his nation's television by American programmes and methods. Will they choose the same sets of words, and connect them together to describe the structure of television broadcasting in the same way? The answer is almost certainly not. Indeed, the coder from Latin America may require words that the coder from the United States does not even think about. So right from the beginning of such cross-cultural research the ideologies inform the basic structural descriptions of the programmes. What price 'intersubjectively shared and verifiable knowledge' now? And do you see why that 'modulo the ideology' is so important to think about?

My concern for such a modulating ideology as Marxism stems not from the human concern it displays, but the claims to truth that many Marxists make, and the rather pitying condescension they display to those who are 'outside the faith'. Clearly, economic relationships are a part and parcel of modern society, and it is easy to see how Marx, in a period of unbridled economic change, saw the root cause of nineteenth-century human misery in the virtually uncontrolled economic forces that were so obvious all around him. If you could change those 'deep structures' then all else would follow, and follow in the sense of

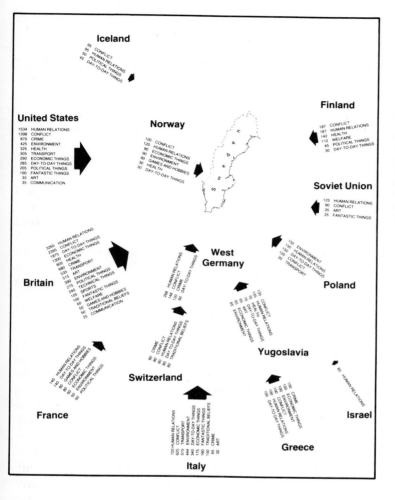

Figure 24.1: Flows of television programmes to Sweden during six weeks in 1977. The subject matter of the programmes was treated in such a way that values were explicitly projected to the viewer. For example, subjects like Crime and Conflict might be the subject of a round table discussion during which the participants displayed moral and ethical concern.

inevitable laws of history at work. To my mind (so again my 'ideology' shapes the discourse), this is understandable in its historical context, but essentially misguided. Misguided, because it invokes a notion of law generated in the physical sciences of things for the human world of conscious, sentient human beings with the ability to reflect on the laws that are purported to characterize them – something physical things cannot do. It is also rather arrogant, because it places the Marxist in a superior, almost detached position 'outside of history looking in', constantly advocating a social progression to a system-wide goal – what a philosopher would call the 'teleological structure' of Marxist theory. Such inevitable 'goal orientation' is once again a child of its nineteenth-century

Darwinian roots in the theory of evolution. We are always children of our time, and an awareness of our own historical position should make us very wary to claim a particular perspective as *the* truth. *A* truth, perhaps, one of many holding for the contingent moment, but not a claim to *the* truth that immediately bounds and limits, and that always means *traps*, the realm of thinking.

Such a claim to *the* truth makes for grave difficulties, not the least because those who are true believers in the Marxist position must find it very difficult to generate any sense of genuine respect for cultures other than the post-revolutionary millennium that is coming in its inevitable lawlike progression. Paradoxically, this is particularly true of attitudes towards Third World countries. If you see history as an inevitable progression (with an occasional bit of back-sliding) from feudalism to capitalism to communism, then you cannot say, in any consistent and non-contradictory way, that you respect the cultural integrity of Burmese, Indians, Nigerians and Jamaicans (or, for that matter, the French, British, Americans and Japanese), allowing them to formulate their own way from the perspective of their own cultural history. Because formulation means choice, but laws of history are iron laws – or not laws at all. The result is that the Marxist ideology, mythologized into a religion, leads directly to the most arrogant sort of Victorian anthropology in which 'we', the enlightened, smile condescendingly at the 'primitive natives' who sleep in darkness, not knowing that the dawn of revolution approaches – until, of course, we have explained it all to them. If you mythologize your truth into *the* truth you cannot help but take a rather arrogant and pitying view of those who do not share your own enlightenment. And, of course, you can have no sense of being trapped yourself, because to think behind the ideology means doubting, and so leaving the very fold where *the* truth lies.

Nevertheless, a number of geographers have met, entered and passed through the claims to Marxist truth, absorbing the concern for moral issues, and the sense of necessity to dig below the superficial appearances. At the same time, they have been deeply bothered by the messianic claims that seem to lead so readily and so often to the sacrifice of human beings today for some promise tomorrow. In the thinking of Marx, individual freedom is too quickly replaced by collective emancipation and salvation in the future. There is also a growing awareness that you usually find what you inquire about and measure. Those pre-structuring ideologies – all of them – lead us to see what we want to see. Few would deny the claims if they were made for *a* truth, but the uncompromising claims to absolute truth are now seen by many for what they are – children of their own time.

Yet I am convinced that these explicit and strong claims have been beneficial to geography as a whole, informing, questioning and enriching our ethical concern and methodological perspectives. They have also sharply heightened our awareness that we all work within frameworks that shape our inquiries, and we always look at problems in partially pre-structured ways. Our thinking about how we think has been deeply affected by a truth contained in the Marxist appeal. In the same way that geography absorbed a quantitative revolution, so it is in the process of absorbing an ideological revolution. As we saw in Chapter 3, poor Geographia always seems to get carried away – dragged from the arms of beautiful Qualifactus by Quantifactus, across the Fluvial Calculus to the arid landscape of

Figure 24.2: Geographia abducted again, this time from an Establishment Quantifactus, across the Rio del Compromiso to the happy Marxian landscape. Compare to the original version, Figure 3.5. Will poor Geographia ever get her own back?

models. But then Quantifactus (Figure 24.2) becomes civilized, puts on a pin-striped suit, carries an Establishment briefcase of approved pattern for the busy jet-set consultant, only to find fair Geographia snatched away by the revolutionary 'El Barbo'. There she goes, poor thing, across the Rio del Compromiso to the land of Marx, the happy peasants gambolling in communal bliss – though with revolutionary work still to be done if the background is any indication! But I have the feeling that Geographia may be getting a bit fed up with being manhandled in such a fashion, and she may get her own back – even yet.

Languages and frameworks: where the structure comes from

25

We have just seen how a particular set of preconceptions about the world shapes a geographer's description, and I think we can understand this intuitively from our own, quite personal experience. We all bring our examined or unexamined assumptions with us when we ask questions about the world, and we realize that the assumptions we make can change our 'way of seeing'. We probably inquire somewhat differently today than we did yesterday, because either thoughtfully or unthinkingly we have slightly changed our views, so the general framework within which we inquire has also altered. In brief, *we* change – as thinking, valuing and growing human beings – and it is always *we* who shape the questions and the answers.

In formal inquiry, where we make great efforts to marshal evidence carefully, and use methods appropriate to the task at hand, there is another 'shaper of descriptions', perhaps one even more subtle, and therefore less thought about. It is language itself, the very thing we think *in*, and I challenge you to try to think without having one of those peculiar 'conversations with yourself' that we call thinking. Try it, and see how your thinking constantly brings a sort of musing dialogue to the surface of your awareness. Even when you visualize geometric figures, graphs and maps, or even algebraic equations, you are still using your natural language to interpret and give meaning to these, to describe to yourself, and perhaps eventually to others, what they signify, and 'how they work'. Those who are genuinely bilingual will confirm that some things sayable, and therefore thinkable, in one language, are difficult, if not impossible, to say and think in another. Which is why poetry, the highest form of language, is impossible to translate perfectly – although a poem in one language may inspire a poet in another to create a wholly new work of art. But this is a rather different, and most subtle, matter.

In scientific work, the 'language' often chosen for empirical description, as well as for theoretical musings, is mathematical, and I have rather deliberately put quotation marks around that word *language* because I still feel a bit uncomfortable saying that various forms of mathematics are actually human languages – like English, Chinese and Greek. This is a terribly difficult question of great subtlety, and not one to tackle in any depth here, but when you hear the word *language* used today – computer languages, graphic languages, algebraic languages, etc. – think very carefully about the way the word 'language' is being used. Having put you on your guard, let us now agree to throw the quotation marks away, and at least acknowledge that any mathematical language we choose to think in may have the same capacity to trap our thinking, confining it to certain thoughts thinkable in the particular structure we have chosen, and preventing us from thinking thoughts 'outside' the linguistic realm we have selected. If you want to go back to the beginning of this paragraph, and read and *think* slowly once again, please do so. I warned you at the beginning of this section (Chapter 24) that some of these questions are not easy for we have come right up to some very new (and yet like many philosophical ideas, very *old*) research frontiers in geography. And geography, it is fair to say, is well in the lead here compared to the other human sciences, where these questions are barely being asked. So let us take our time, and chew these ideas slowly and carefully.

Until roughly a hundred years ago, a great deal of mathematics was inspired by attempts to describe faithfully the physical world. Even such simple things as counting with the integers, and doing operations like adding and subtracting, have very practical descriptive purposes, because they provide answers to those simple 'how many' questions that inspire the large realm of *quantitative* mathematics. As for geometry, it is nice to have formal proofs of theorems like Pythagoras', but 'the square on the hypotenuse is equal to the sum of the squares on the other two sides' was known to the Babylonians, Egyptians, Indians and Chinese, and used by them for the thoroughly practical tasks of surveying and describing the sizes of fields, and making sure the corners of temples and pyramids were square. Much later, Isaac Newton developed his 'infintensimals' to describe the movements of the moon and the planets, and with Gottlieb Leibniz started a branch of mathematics we call the calculus today. Karl Friedrich Gauss – the Prince of Mathematicians – developed the distribution that bears his name (the normal, or Gaussian, distribution in statistics) to minimize the errors in his surveying instruments when he was making maps for the Duke of Hanover. Astronomers like William Hamilton even developed new sorts of numbers that were needed for a description of the world, and invented, with his friend George Boole, forms of mathematics that were quite outrageous, forms where a $\times$ b does not usually equal b $\times$ a – although it can sometimes! Today this sort of mathematics is used routinely by teenagers in school, and we call it matrix algebra. The same tradition continues: physicists like Roger Penrose invent new algebras to describe the strange behaviours of subatomic particles at the quantum level, and at the other end of the scale, the cosmological realm of the universe, strange but useful calculi are created out of a relative newcomer to mathematics – topology. I cannot help pointing out that *topos* is the Greek word for *place*, while *logos* can mean *knowledge*, so topos-logos, knowledge of place, comes close to geography once again. We know the links between geography and mathematics were very strong at one time, so we should not be surprised if we find certain types of topology being used in geography today.

In this brief and highly compressed 'practical history' of mathematics there is an important lesson to be learnt. In general, those who have worked in the physical sciences have created the mathematics they needed to describe what they saw and wanted to understand in the physical world. A very deep study and careful reflection on the things themselves *preceded* the creation of the sorts of mathematics that were needed to describe them faithfully – by which I mean 'without doing too much distorting (unfaithful) damage to them'. Of course, there were also occasions when physical scientists saw from their study of things that they needed a 'different sort' of mathematics from the usual ones to do the job required, and it just happened a form of mathematics already existed that was just what they needed. Crystallographers wanted to describe symmetrical properties and quickly latched on to group theory; Albert Einstein needed things that could describe lots of dimensions at once and found the tensor calculus very handy; and the great physicist Walter Heisenberg was trying to invent some strange mathematical operations for tables, until Max Born told him to go off and study matrix algebra – already well developed. Even so, these are not really exceptions: the *need* for a mathematics with particular properties still came from intense study and careful reflection on the things themselves. Then, and only then, were the forms of the required mathematics created for the descriptive task.

What sort of tasks were these? Almost without exception, they were tasks of *mechanical* description – celestial mechanics (Newton); statistical mechanics (Boltzmann); electro-mechanics (Maxwell); quantum mechanics, continuum mechanics, and so on down to just plain old-fashioned *mechanics*. Faced, over and over again, with the task of describing a mechanical world of things, the physical scientists set an impeccable example of creating all sorts of mathematical languages for the express purpose of describing mechanical things. The physical world is described as a mechanistic world, and its mathematical languages are designed right from the beginning to capture its mechanical nature. It is a world in which we try to predict and control, in which we try to foresee, in which we try to look ahead to tomorrow to find the consequences of today. In other words, given the state of the system at time t_0, what will it be at time t_1? Too formal, jargony and symbol-ridden? Not really: just think why you listen to the weather *forecast*. Huge computer models, made up of tens of thousands of equations, try to work out from the state today (t_0) what the state of the weather will be tomorrow (t_1) through the *mechanisms* involved. The physical sciences (of which meteorology is one) can use mathematics in this way because they are, in an extraordinarily deep sense, *already mathematical*. For the ancient Greeks, *ta mathemata* meant 'that which man knows in advance in his observation of whatever is and in his intercourse with things', and in quoting Martin Heidegger, perhaps the greatest philosopher of this century, we must remember he was a close personal, and professionally intimate, friend of Walter Heisenberg, who considered Heidegger one of his two or three peers. The physical world of unconscious, non-sentient, and unreflective things is, in our human eyes (and who else's after all?), a mathematical world. Above all, it is a world whose essence is captured by the function.

Now I do not want to introduce the symbolism of particular forms of mathematics – this is not the time and place, and specifics become much too complex for the sort of general consideration that we are thinking about here – but the basic and important idea is very simple. The function is nothing more than $Y = f(X)$, which says that something we call Y (which can be anything we like, perhaps yields of wheat, average longevity, etc.), depends on something we label X (rainfall, daily alcohol consumption, etc.). More specifically, we might write $Y = a + b \cdot X$, and if $a = 1$ and $b = \frac{1}{2}$ we can be even more specific and write $Y = 1 + \frac{1}{2} \cdot X$ (Figure 25.1). We can even make a graph of this: when $X = 0$, $Y = 1 + \frac{1}{2} \cdot 0 = 1$;

$$y = f(x)$$
$$y = a + b.x$$
$$y = 1 + \tfrac{1}{2}.x$$

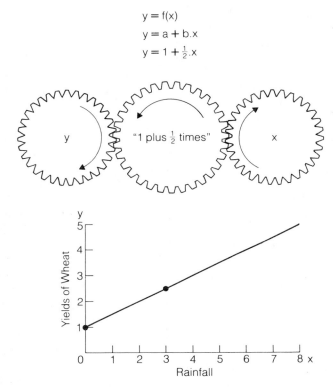

Figure 25.1: The general function Y= f(X) specified as a *linear* function Y= a+b.X, and graphed as the specific linear function Y=1+½.X, where Y is yields of wheat, and X is rainfall.

when X = 3, Y = 1 + ½ .3 = 2 ½ . . . and so on. But notice that X is like a cogwheel connected to the cogwheel Y through those signs of multiplication. and addition +, things a mathematician would call *binary operations*, because they work (*opera*) on two (*binary*) things at a time. Every time cogwheel X clicks around, so does mechanically coupled Y. This means that for every value of X, there is one, and only one, corresponding value of Y; each X in a set of Xs is linked to a Y in a set of Ys and *vice versa*. This means that if we know the rainfall we know the wheat yields, or if we know what the yields of wheat were that year we could work out how much rain fell. As for longevity (Y) and alcohol consumption (X), things are a bit more complicated (Figure 25.2), because the function mechanically coupling the two together is not a straight line (a *linear* function), but a curved one called, in this case, a quadratic. This says that the longevity of teetotallers (X = 0) is less than those who imbibe a moderate amount of alcohol each day. My grandmother used to say, ' little bit of what you fancy does you good.' Today we replace such an old wives' aphorism by the scientific statement that alcohol tends to lower

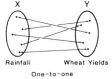

307

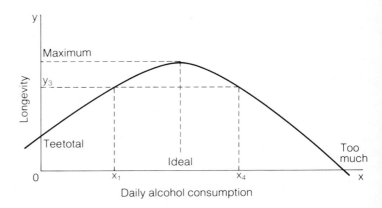

Figure 25.2: The function Y=f(X) now specified as a *quadratic* function Y= a+b.X−c.X², where Y is average longevity and X is daily alcohol consumption.

the level of low-density lipids that build up cholesterol. As a philosophical position, I prefer my grandmother, and Freudians can make what they like out of that one. Notice, however, that too much of the stuff and your liver starts to give out and your longevity declines. As always, the Greek ideal of moderation in all things.

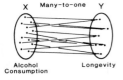

Our little picture of the function, as a set of Xs connected to a set of Ys, also changes, because now there are always two Xs going to every Y (except for the peak of the curve where just one value of X is involved). The mechanical coupling is also still there if we go from the Xs to the Ys, although it is a bit more complicated. If we know the alcohol consumptions are X_1 and X_4 we know the longevity is Y_3. The problem comes when we try to go the other way: given the longevity of someone, say Y_3, how can we tell whether their alcohol consumption is X_1 or X_4? Are they a moderate drinker, who could increase longevity by imbibing just a bit more? Or someone who is overdoing it, and really ought to drink less?

Whether wheat yields and rainfall, or longevity and alcohol, this discussion of connecting Xs and Ys together with binary operations is really at the heart of the question of mathematical languages. Without going deeply into details, and at the risk of making my mathematical friends apoplectic, we can point to two very important ideas. First, when every X in a set is connected to a Y, and *only* one Y (as they were in both our linear and quadratic examples), we have what a mathematician calls a *function*, and that, as we have seen, is the sort of mechanical coupling that characterizes the world of physical things. Functions always take one, or perhaps many, Xs and connect them with just one Y. This is why a mathematician calls them a one-to-one, or a many-to-one *mapping*, and once again we see how geographers and mathematicians are

308

connected together by their language. But even in the case of our quadratic, let alone more complicated functions, we saw that we could not 'get back', because our Ys were not connected one-to-one (or many-to-one) with our Xs. A mathematician would say that going from Y to X we have a one-to-many *mapping* (even when, as in this case, 'many' is only two). So a mapping is a much less constrained way of connecting things together than a function, and there is no reason why we cannot go all the way and have a many-to-many mapping. So we see that all linear functions are functions, but not *vice versa*, and all functions are mappings, but not *vice versa*. It is another example of all dogs are animals, but not all animals are dogs. But we can go even further: in our examples of connecting elements of sets (to use the official terminology), we had to use up all the Xs (although not, strctly speaking, all the Ys). If we now release this last constraint and condition on our ability to describe connections between elements of sets, we have what we call a *relation*. So now all functions are mappings are relations, but not *vice versa*.

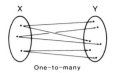

One-to-many

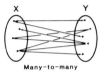

Many-to-many

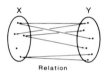

Relation

So is it not strange that when people in the human sciences approach the world, and try to describe it in mathematical ways, they invariably look at it through the small porthole called the function, and even constrict their vision down to the tiny keyhole of the linear function, when all they have to do is throw open wide the great door of the relation? Of course, when you throw open doors, things you saw through the keyhole may look quite different. As a scientific voyeur, you may decide after all that you like the Peeping Tom view better, because it is more familiar and easier to understand. But if you throw open the door, and start with the relation, and it turns out that the functional keyhole is appropriate for the descriptive task (as it undoubtedly is over much of the physical sciences), then you will always find the functional form. Remember, all functions are relations. On the other hand, if you start by restricting yourself to the functional keyhole, you will never know what the world looks like in a freer, less constrained language. And since we think in language, what does this say about your thinking?

So what have all human scientists, geographers among them, invariably done? Well, one thing is certain, they have *not* followed the impeccable example of the physical scientists, carefully investigating and reflecting upon the things that have caught their attention, the things they intend to understand. Invariably, they have borrowed unthinkingly the forms of mathematics devised to describe a mechanical world, and forced the

309

rich connective tissue of the human world on to these highly constrained forms. But since these forms were created to describe a mechanical world, the human world described in these ways can only look mechanical. The language chosen does not allow it to look like anything else. Do you see now why I was a bit concerned about geography as a child of its technological time in Chapter 4? This relational view opens up thinking once again to freer, less constrained, and essentially *qualitative* forms of mathematics. People are always complaining that you cannot put a number on everything, that some things cannot be measured. They are quite right: quantitative mathematics is a much more highly constrained form than qualitative mathematics, and it is always restricted to sets of numbers and operations on these sets. In contrast, qualitative mathematics can choose any elements it likes, and connect them in any way it chooses, and most mathematics today, contrary to the impression we get in school, is qualitative, with not a number in sight.

But there is a second, very important thing to point to in this discussion of sets and relations and the mathematical languages that can be constructed out of them. You may have noticed that the ordinary language of this book is permeated with the word *structure*, and this was also one of the key words that exploded in the publications of geography (Chapter 3). The reason is obvious: our everyday discussions, including the geographic, are also riddled with the concept of structure. We talk about the *structure* of society, international trade, a game of chess, a poem, a ballet, a molecule . . . the list is endless. As humans we are fascinated by structure, because structures tell us how things are connected together. And how things are connected together means relations on and between sets of elements. A set of elements (dancers, words, amino acids, chessmen, people, etc., etc.) is just . . . well, just a set, nothing more – no connections, no order, no *structure*. As for a relation, you cannot have a relation without some things, some elements, to relate. But put elements and relations together, and structure appears, and is it not *that* which we intend to understand? Even the influence of the Marxist perspective is generally labelled *structural* Marxism.

What we have seen in geography is a gradual expansion of mathematical languages, and with it a deeper and more thoughtful concern for that truly basic idea of structure – including the structures of the mathematical languages themselves, the languages that allow us to describe structural properties of irrigation schemes, the cognitive maps of handicapped people, the ways taxes are collected and dispersed, the flows of migrants from Swiss villages, television programmes, and so on: here, too, the list is almost endless. For all its mathematical formality, it is a very exciting intellectual story, with much experimentation, exploration and building outwards from the first borrowings. Michael Dacey, for example, investigated the degree to which some fundamental mathematical properties of geographic distributions might disclose the processes that could have led to the particular patterns, and later focused upon the highly formal statements (called axioms) from which we might derive systems of central places. In the process, he contributed to mathematics itself, unknowingly following the old Arab tradition of al-Khorizmi. Michael Webber, in a now-classic study of pioneer settlement, stated in ordinary language the basic concepts underlying the settlement process, translated these into the continuous mathematics of the calculus, and derived an extraordinarily close description of the way the towns developed in Iowa, and how the system 'shook itself down' to produce the patterns we see today. Employing a similar approach, but focusing upon the broad and essentially *qualitative* aspects of the mathematical description, Jorgo Papageorgiou has gone inside the city, asking how very general attributes interact to produce the variety of patterns we see in an urban landscape. Every city is different, but perhaps all the forms in the urban kaleidoscope are just different outcomes of the same deep basic forces at work?

Michael Webber
Macmaster University
1941–

Geographic descriptions in other mathematical languages also seem to point to the same idea, and computer languages are becoming particularly prominent. Whatever the computer language (and there are many today), it always makes provision for carrying out certain operations on things, and it is these sequences of commands that lead to what we might call the algorithmic approach. As we know (Chapter 2), *algorithm* is an old mathematical-geographical idea, involving breaking down an often long process into small, easily understood steps, and then chaining them all together. To the degree that the separate pieces and their linking sequences represent and reproduce what people do, we can 'write' our models, and get computers to work through the consequences. This is really the approach of the field called artificial intelligence (AI), an area where Terrence Smith has been particularly prominent in the modelling of human behaviour, particularly that sort of behaviour characterizing the *searching* of geographic space. How *do* children find their way? Can we understand, and write down in formal ways, the sequences of trials and errors they perform,

Jorgo Papageorgiou
Macmaster University
1936–

Terrence Smith
*University of California,
Santa Barbara*
1943–

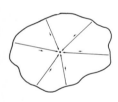

Helen Couclelis
*University of California,
Santa Barbara*
1944–

and does such an algorithm help us understand how handicapped children have great difficulty 'coming to grips' with an urban neighbourhood, or why they fail totally to navigate it? Can the search behaviour of people buying houses in an urban area be modelled in the same algorithmic way? These are exciting questions, on the very frontiers of the human sciences, and once again they point to the fact that very general processes may lead to individually different results – children learning to find their way, and adults searching for housing, display a variety of possibilities, not a single outcome.

This idea, that similar processes may work to produce different superficial appearances (using *superficial* in its true sense of 'lying at the surface'), has been a concern of Bernard Marchand, who we have met a number of times before, and another theoretician, Helen Couclelis. Both are concerned with highly abstract, geometric properties of urban structure, with Marchand calling attention to the effect of designating a particular spot as the site of initial settlement, the effect of a privileged point in the urban geometry that controls so much of the form in the future. Working in particular with Los Angeles, but drawing from his intimate knowledge, experience and love of his own Paris, he has approached the description of urban form from many different directions, ranging from the abstract mathematical, through psychiatric (particularly Freudian) interpretation, to the philosophical (Hegel's dialectical view of processes creating their own contradictions). In contrast, Helen Couclelis has called attention to the very deep geometrical properties and structures that are prior to any actual development of a city, including something so apparently simple as the effect of bounding a space and treating it as a set of finite points – points we might be able to observe and count. On an infinite plane no point is more accessible than another, for the sense of a central point cannot be defined. Only when a boundary is put around a piece of the space, does one point become more accessible than another. In a sense, the view of Marchand designates the point of original settlement, and lets the city grow and define its own bounds; while the approach of Couclelis bounds the space, and so produces a privileged point.

Helen Couclelis has also brought explicitly into view the question of what happens when geographers start to aggregate their data, and move from the finest grain view of an urban system, recording all the details of each physical thing and human being that goes to make up a modern city, to the most general view, where only a bare and totally abstract geometry

312

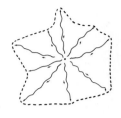

is left. This question of aggregation is a thorn in the side of every scientist, particularly the geographer who has to deal not just with the behaviour of the human actors, but also take into explicit account the spatial properties of the geographic stage upon which the poor players strut. We have to work with two hierarchies of aggregation simultaneously (Figure 25.3), one describing our view of the human beings, the other the geographic space.

At the lowest (or is it the highest?) level (7), we have fully conscious, real people making real decisions. In principle, we might be able to interview each person in a town in any degree of depth we require, posing a series of questions which are answered Yes (1) or No (0). In fact, much of the information would be quite useless (did you brush your teeth today?), or

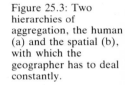

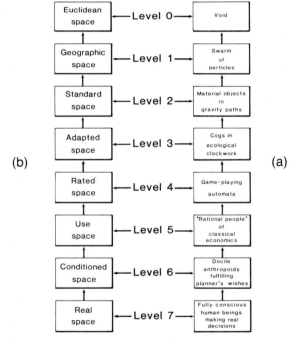

Figure 25.3: Two hierarchies of aggregation, the human (a) and the spatial (b), with which the geographer has to deal constantly.

unobtainable (did you dream of crocodiles last night? – sorry, can't remember), or downright impertinent (how *dare* you!). So we might select responses to certain questions and not others, describing each person as a string of 1s and 0s, a string as long as is required for the appropriate description. Corresponding to these detailed binary strings representing people, we have real space, with all its streets, pavements, traffic lights, buildings . . . and so on, as it is actually perceived and evaluated,

313

described again in any level of detail we require. In practice, of course, we could never specify *everything* about the people and the space they perceive themselves to be in; and even if we could, we could never handle all that information, most of it quite irrelevant anyway, even with the largest computers. So we abstract, and that means we essentially take a sort of conceptual sieve, and sieve out the bits about people and places we do not need. Arriving at level 6, we have docile anthropoids who are assumed to carry out the wishes of the planner, and a conditioned space, conditioned in the sense that it is determined by institutional and social determinants, the shared collectivities of perceptions and evaluations, rather than those thousands of individuals. As we apply our conceptual sieve again and again, we reach the 'rational people' of classical economics, the cogs in the social clockwork, swarms of particles subject to gravity fields (remember the gravity model?), until everything is sieved away, even the probable positions of the point-particles which our people have now become. On the spatial side, we have sieved away more and more characteristics of the space, its utility to people, its value, land uses, shapes, locations, boundaries . . . until we are left with the purest of geometrical properties – in this case Euclidean space has been chosen, simply because it is familiar, but other geometries are possible.

Thus, we have different perspectives of abstraction on both the players and their stage: both are sieved and aggregated away as we move up the Couclelis hierarchies, from impossible-to-deal-with reality at level 7, to such levels of total abstraction that our final result is a totally banal void at level 0. Yet somewhere in between, whether as an ordinary citizen trying to 'make sense' out of pieces of the puzzle, or as a professional geographer trying to inquire formally into the complexity, somewhere in between is where we all work and think (Figure 25.4). And notice two things: firstly, that hierarchy is *not* the usual one in which something at one level has to aggregate to one thing only at the next level up. We are too used to thinking about hierarchies in terms of curate-parson-bishop-archbishop (a *hierarkhes* – *hieros* sacred + *arkhes* ruler – so now we see where *hierarchy* comes from); or private-corporal-sergeant . . . general; or the typical organization chart of a company, in which one assistant manager reports to just one manager above. Of course, we all know such organization charts are often worthless, because in order to do a job properly a person may have to report to three people above, and somehow coordinate their requests. So it is in our hierarchies here: a person at the lowest level may aggregate to the set of females, wives, bus
314

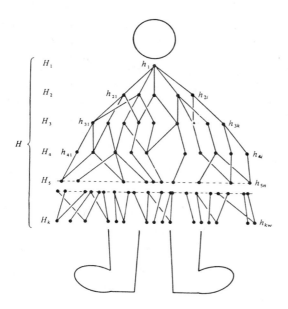

Figure 25.4: The 'Urban Man' of Helen Couclelis, made up of hierarchical levels joined by complex relations of aggregation.

drivers, mothers, at the next level of social roles, and so on up the line. So we see now that aggregating relations may be one-to-one, one-to-many, or many-to-one mappings. However, in general, we lose information as we sieve away, and that means that those binary strings of yes-no's representing our people at the lowest level are constrained down further and further, perhaps through some complicated sequence of relation to mapping to function. Perhaps *this* is why the functional mathematics borrowed from the physical scientist sometimes 'works' reasonably well in highly 'aggregated' geography.

Secondly, I also have the feeling that the implications of this aggregation process by mappings go even further. When we aggregate people, when we sieve away their individuality, we scrunch them, and *scrunch* comes from *scrag* (to wring the neck) and *crunch*. So we have to be careful here: scrunching people with hierarchical mappings is all very well, and squeezing them into mathematical models, where the binary operations treat the people-lumps as *things*, may help us understand some overall process at an abstract level, but what happens when we go the other way? These mathematical models are being used more and more today to plan and shape the world we live in. What happens when we 'unmap', when we *dis*-aggregate the conclusions of our models in a series of policies? What do these then *mean* to all those poor binary strings at the bottom of the hierarchy? Is this why bureaucracies are so dangerous, and so generally loathed by people who are

315

not the 'docile anthropoids fulfilling the planner's wishes'? Too many bureacuracies, originally developed with impeccable human intentions, become self-sustaining, self-protecting, and ultimately self-serving people crunchers.

If we acknowledge that all the mathematical descriptions scrunch people, why go in that direction at all? Why not stick to plain old ordinary English (or your natural native tongue), and provide the rest of us with deeply penetrating descriptions that illuminate a portion of our human geographic world? It is a good question, one often asked by people who, quite paradoxically, take either an artistic or a brisk no-nonsense view of the world. But it forgets two things: firstly, the enormous ambiguity and imprecision of ordinary words, many of which are at such high levels in a 'hierarchy of meaning' that they can mean all things to all people. Think of words like 'socialist', 'freedom', 'liberty', 'democracy', words beloved by politicans because people can put whatever meaning they like on them (until after the election, when they realize in actual deeds what the politicians meant). Think, too, how words like these take on diametrically opposite meanings under certain political regimes, as George Orwell saw so clearly. At one level, we can rejoice in ambiguity, and both poetry and prose would be destitute without it. Here the play on words, the double and multiple meaning, is used constructively, artistically and, not the least, *ethically*. But in a scientific description, where we are trying to create that 'intersubjectively shared knowledge', we have to tighten the net, and catch with a fine mesh of definitions the slippery silver fish of words. Few politicians want to be pinned down: people might begin to understand what is being said, instead of hanging their own wishful thinking on their verbal pegs. For scientists, clear and open definitions are essential.

The second thing we forget is that ordinary language has its own structure, and produces its own constraints. And since we *are* our thinking, and thinking *is* in language, so we are ultimately constrained once again. At this point, as James Joyce and other poets knew, there are only two things to do: increase the set of words (elements) you can employ, and loosen the rules (relations) that link them together. It is a dangerous human game, but sometimes the only way to go. It is the way chosen by Gunnar Olsson, perhaps a way of last resort, a way out after a long apprenticeship spent on a thorough examination of formal mathematical models, and a meticulous examination of formal logics as languages adequate to the descriptive task faced by human geographers. In both cases, models and

Gunnar Olsson
Nordic Institute for Studies in Urban and Regional Planning
1935–

logics, his answer was ultimately No, and following the example of his hero Joyce he decided to describe, interpret and comment upon the text of the world around him, constantly bumping his head against the limits of language.

It was a remarkable decision and feat, for the language was English, and Olsson is Swedish. The ability to use another language creatively is rare – Conrad (Polish-English), Beckett (English-French), Nabokov (Russian-English), Kosinski (Polish-English) come to mind – and demands enormous efforts, concentration and, above all, caring. It was a lonely and misunderstood road, sometimes excoriated and dismissed, particularly by English reviewers. Isolated and cut off from deeper continental traditions of thought, while assuming the Continent was cut off from them, an intellectually older English generation has difficulty in understanding the sense of the constraining question, for it seems to lack a string on the intellectual violin tuned to resonate with the concern and caring. Meanwhile, a newer generation is appearing that senses both the linguistic and social questions of personal freedom and systemic constraint. They are not alone, and if contemporary geography has one outstanding characteristic today it is the sense of reaching out, and linking up, in common concern with other thoughtful and reflective men and women along other frontiers of the human sciences. It is to that renewed tradition of geographic reflection that we must now turn.

26 Geographic reflection: renewing an old tradition

It is no accident that two words in the title of this chapter start with *re-*, that old prefix from the Latin meaning *again* and *back*. To *reflect* means 'to go back in thought . . . or consult with oneself', a lovely dictionary definition of those conversations with yourself that we call thinking. To *renew* means to 'begin . . . anew, continue after intermission', which says that something dropped is picked up and carried forward once again. Let us see how thinking, discarded and forgotten at one time, is being taken up again in geography.

Not many years ago, perhaps a mere 300, we could not speak of physics, chemistry and the other physical sciences in the way we do today. In the universities there were schools of medicine, jurisprudence (law), theology and philosophy, the first regulating the body, the second ordering the larger society, and the third taking care of the spirit. All other questions were posed and answered by philosophy, providing the answers did not conflict with theology – as Galileo quickly found out. Then, in the seventeenth century, certain areas of inquiry started to break away from the old holistic field of philosophy, terming themselves first 'natural philosophy', and then splitting further into physics and chemistry, until today we have a broad spectrum of highly specialized fields investigating different aspects of the physical world. In the eighteenth century, we see other subjects hiving off, this time as the fields of inquiry into the living world start to break away, and today we find a second array of specialized lines of inquiry from paleobotany to cytogenetics. In the nineteenth century the process of fission continued, and for the first time the human sciences like political economy, human geography, psychology, anthropology and sociology appear. The dates are not exact, but will serve us well enough to point to main trends.

The problem was that in detaching themselves from philosophy, the new fields gradually left behind the truly deep and outstanding characteristic of philosophy itself: a tradition of

318

careful reflection and thinking about those who sought the truth, the morality of the search, and the truth itself. In fact, the new sciences of physics and chemistry made so much progress opening up and explaining the physical world that they arrogantly elevated themselves to the finders and guardians of truth. In the process, they discarded what they condescendingly termed mere 'metaphysics', and claimed that their 'scientific method', characterized by the purest logic, was *the* way to *the* truth. As for moral and ethical questions, these too would be resolved by purely formal logical argument. After all, it was by being logical, rather than merely emotional, that we were distinguished from the lower animals. In occasional moments of rare modesty, it had to be admitted that the truth was open, but few doubted the basic picture of diligent workers marching confidently along a broad highway of Science to a great shining Truth beckoning at the other end. Unfortunately, few realized the still-deep and unspoken theological implications of such a view, mainly because so few thought 'philosophically' any more. By the 1870s, many physicists literally thought it was a matter of a mopping up operation here and there, a tidying up of a few loose ends, but essentially the basic picture of the physical world was complete.

We all know what happened next: Michelson and Morley, Planck, Einstein, Heisenberg, Bohr . . . the physical parcel, so neatly tied up, exploded, and the quarks and gluons, with all their charms and colours, are still raining down. Significantly, the tradition of reflective thinking in the physical sciences was taken up again, or perhaps was never entirely dropped, by some of the best physical scientists, who began to see that the physical world 'out there' is what we make of it, and what we make of it depends on where we stand historically, and on the mathematical languages we choose to create to describe it. There is also the realization today that the traditional approach of the sciences, which takes everything apart into smaller and smaller pieces, does *not* necessarily help us understand how the larger assemblages work. Knowing the structure of the DNA molecule does not help us one bit to understand the organization of an ant colony, or how bees communicate, let alone biological and human systems orders of magnitude more complex.

Any field of human inquiry is impoverished and diminished by discarding that older tradition of reflective thinking, and nowhere is the loss seen more clearly than in the human sciences, where intentional inquiry is made into the many facets of the human condition itself. All the human sciences have paid a severe price for cutting that old connection with philosophy, and, once again, to point to such a severing is more than just a polite genuflection to the Greek tradition that runs like a slender golden thread through our Western world. Because if that tradition is irrelevant, we have to ask why the dialogues of Socrates cannot be taught in Prague today, why teachers who do so are dismissed from their university posts, why secret police hound the seminars in private homes, and why some are forced into exile. Thinking, even very old thinking, is a dangerous and difficult activity.

It is, nevertheless, an essential activity, with the obligation always to keep thinking open, to struggle, again and again, with the same questions as they appear in different historical circumstances. These questions are the old ones of right and wrong, of the individual and the society, of the circumstances of our own knowing. They are ridden with tension by the constant struggle to find a point of reflection 'from the outside looking in', that Archimedean point from which we can see and reflect upon ourselves, even as we acknowledge that we can never stand wholly outside, and that the points we find are given to us by the particular, and ever-changing, historical circumstances in which we find ourselves. Yet it is the tension between the desire to get outside, and the wry, perhaps self-mocking acknowledgment that we can never do so, that keeps such thinking open. Otherwise, and perhaps sometimes out of sheer intellectual exhaustion, we declare our faith and allegiance to *a* way of seeing, and cling to *a* truth that quickly closes down and becomes mythologized into *the* truth. Then we are trapped, except that our thinking is now such a mixture of arrogant certitude and flabby self-righteousness that we never even feel the jaws cutting into the mythologized jelly we have created.

It is this constant willingness to create and bear the tension of critical thinking, to acknowledge the trap of one's own time and place and circumstance, yet feel the steel jaws cutting as you struggle to escape, that characterizes the best efforts in geography today to move back to that older, discarded tradition. In the process, geographers are meeting others from the splintered and disconnected fields that the human sciences have become, so there is beginning to emerge, once again, a shared vision of a more integrated, holistic discipline. I must not exaggerate: most human scientists, geographers included, continue their day-to-day inquiries in the accepted, 'normal', i.e. unthinking way. The vanguard of one era becomes the rearguard of the next, and as the philosopher Hegel noted in his own day, ideas begin life as liberating forces and evolve in the course of time into suffocating straitjackets. The strings on the conceptual parcels must never be tied too tightly: they always have a way of coming undone.

In this renewed tradition of open thinking, a basic question posed by Allen Scott is, why do geographers study what they do and when they do? Why, historically 'now' (1955 to the present), do we find a geographic (r)evolution? Why do we see the words *planning* and *region* and *structure* exploding in the literature (Chapter 3)? Why the influence of the Marxist perspective informing geographic inquiry (Chapter 24)? Without denying certain degrees of intellectual independence, all knowledge and science are strongly influenced by the historical circumstances in which they are found. As beings in time, we are always children of our time. Perhaps, then, we should not be too surprised to find that much of the content of modern geography is given by the problems of what has been called the 'late Capitalist society', where all the human relations – social, political and economic – are eventually projected through geographic space.

Allen Scott
*University of California,
Los Angeles*
1938–

It is important, as always, to use language carefully here, and not generate unthinking reactions by using such words as 'capitalist' carelessly. In virtually all societies today, production is essentially managed by the state. In the post-Keynesian world of the welfare state, enormous resources are devoted to the control and management of *all* societies to prevent the ups and down of economic crises, to regulate the business cycle, and to manage the human and social dislocations that such changes bring. Regulation of working hours and conditions, of pollution standards, of foreign exchange, of taxes . . . all are stated and enforced in law. Huge amounts of money are devoted by the state to such things as road-building, communications, urban renewal, regional development and public housing, and when large corporations like Lockheed and Chrysler are about to go under, because of the 'free interplay of market forces' (?), subsidies and loans of hundreds of millions of dollars are found to prevent the social, economic and political dislocation that such closings would bring in their wake. 'Late capitalist society' is a managed and controlled society in which social stability is insured by bureaucratic intervention. To take only one of many examples, in France the choice is quite explicit, and was made long ago, no matter what the political party in power. France is managed almost exclusively today by those who have demonstrated their ability in *mathematics*, for this is the central examination that constitutes the gateway to the *hautes écoles*, from which flow streams of highly selected engineers and econometricians, all well versed in control theory informing the model-building process going on high up in the people-scrunching hierarchy (Chapter 25).

Such highly managed societies have evolved slowly as alternatives to the humanly distraught conditions of the nineteenth century, and we see today constant intervention to secure economic and social stability. What we should note here is that these societies produce a highly problematical geography as the technical 'solutions' conflict with the management of human relations and the reproduction of society itself. For example, highways designed to relieve bottlenecks and increase circulation destroy large areas of poorer neighbourhoods in cities. But as America learnt in the 1960s and 1970s, the poor are not so manageable today. When the organization of geographic space changes, and produces discordant effects, the geographic constraints themselves have to be managed, planned and controlled. It is worth noting that Scott comes to this task of reflecting deeply on why we think as we do from a background of highly formal and difficult combinatorial programming. Emerging from such a background, he has pointed to the need for constant vigilance and a high level of analytical self-consciousness. Not only do geographers have to see beyond the technical and methodological minuets that are being danced, but even as they move to new discussions they have to generate the capacity to see beyond these in turn. Always that Archimedean point moves, like a beckoning will-o'-the-wisp, to think beyond the current frontiers – open frontiers today that become tomorrow's constraining bounds on fresh, and ever-renewed, thinking.

It is this sense of seeing geography socially embedded in its own times that has produced some of the most exciting and reflective thought, and created a strong drive to link geography with developing social theory. The renewed connections represent a two-way street: the thinking of social theorists like Anthony Giddens infuse geography with a heightened sense of awareness, but the geographic insistence that all social action is played out in *space* as well as in time is infusing, informing and enriching social theory in turn. As Allen Pred has pointed out, it is no accident that social theorists have seen in the paths, bundles and projects of Torsten Hägerstrand's space-time dancers the crucial importance of the organization, patterning and structure of geographic space, the stage upon which all the players undertake their always constrained movements.

Anthony Giddens
University of Cambridge

A prominent feature of critical social theory today is a radical rethinking and reinterpretation of geographic space and historical time, and again it is not a matter of tacking on geography to social theory, or social theory to geography. Edward Soja has pointed out that society cannot be understood divorced from the geographic stage, any more than the organization and structure of geographic space can be understood if it is torn from the context of the social forces at work. His insistence upon infusing social thinking with what he terms the 'spatiality' of the contemporary geographic outlook takes up that spatial component we saw exploding in the late 1950s, and renews that concern at a time when many of the traditional Marxist positions are ready to discard it as a mere distraction, unsanctified by Marx or Lenin, neither of whom were distinguished by their geographic insight or awareness. Perhaps the outstanding characteristic of geographers engaged in the construction of critical social theory is the sense in which one feels they have met, passed through, and appeared on the other side of the Marxist perspective, absorbing that which is of continuing and humane value, but refusing to become mythologized jellies themselves.

Allan Pred
*University of California,
Berkeley*
1936–

An additional benefit, as geography and social theory renew old ties, is that we become more acutely aware of some of the forgotten connections between geographic models and the social conditions in which they were constructed. We see today how early land-use models actually emerged from discussions of social relations between landlords and peasants, how location theories were part and parcel of the response to the early effects of industrialization, and how some basic urban models emerged from an early concern for social relations in a highly dynamic urban context of late nineteenth-century America. Always the individual is born into a pre-existing society, thrown into a world not of his or her own making, caught up in spatial and social structures that allow and forbid. But, as Derek Gregory (to whom we owe many of these historical insights) has pointed out, in the spatial and social 'prisons' we are not in solitary confinement. In brief, we are all in the same boat, even if we are not always fond of some of our fellow passengers.

Edward Soja
*University of California,
Los Angeles*
1940–

The prisons we find ourselves in are actually the social and spatial *structures*, and once again we find that word permeating the discourse. The problem is that usually the word 'structure', so intuitively valid and useful, is seldom defined and made truly operational. Once, in a mood the French would call *méchant* (a sort of cross between the English *mischievous* and *bloody-minded*), I counted the number of times *structure* was used in a short article written by some geographers concerned with the interplay of social and spatial dynamics in an American city.

Derek Gregory
University of Cambridge

323

Thinking about what we think

The total came to 75, but it was always used as a high-level buzz word, and you wondered what the answer would have been if you had asked in wide-eyed innocence: what do you *mean* by structure? How would you define it? How, having defined it, would you make it operational, capable of being truly investigated? Social theory, like all theory, eventually has to be grounded in real things (elements of sets) and real connections (relations on and between sets). Otherwise such theory never fully informs empirical inquiry, nor is it informed by such inquiry in turn. I have the strong suspicion that when the disciplined approach of the new mathematical concern for structure meets the deeply human and social concern for structure that we are going to witness an intellectual fusion of great excitement and worth.

We are also reaching a rather critical junction point, where the very best, hardest and clearest thinking has got to inform geographic inquiry in the future. In approaching something as complex and as delicate as human society in geographic space, with all the modern theoretical, methodological and technical insights now at the disposal of modern geographers, we need, as perhaps we never needed before, that critical self-reflective stance that the tradition of philosophy once gave to the areas of inquiry that we call today the human sciences. So within the context of three perspectives posed by the German philosopher Jürgen Habermas, perspectives we shall use as simplifying pegs on which to hang a discussion of a very complex process, let us think what happens when a geographer inquires into one of the many juxtapositions and combinations of human society, geographic space and historical time.

At the beginning of any inquiry, there is always a choice to be made, the choice of what shall be the things and relations forming the subject of inquiry itself. Shall we focus upon the ethnoscientific knowledge of Indian farmers? A particular historical record of technological change? Recent rural to urban migration in a Third World country? Highly abstract or geometrical propositions about urban form? Whatever the focus of inquiry, choices have to be made that some things are relevant, while others are not. Those choices about the sets of things we want to observe, and about the various relations that structure the sets, are fundamental, for they shape all the subsequent investigation. Then, after the definitions have been stated, the methodological choices have to be made, always sensitive to the fact that these also trap and shape the inquiry, even as informed inquiry cannot go forward without clearly stated procedures. Next comes the choice of language, and the

Jürgen Habermas
Johann Wolfgang Goethe University, Frankfurt
1929–

324

awareness once again of how such a choice shapes what can and cannot be thought. All of these things, however they manifest themselves in definitional choices, ways of collecting data, mathematical and computer languages, the making of maps, the construction of models . . . all these things are essentially aspects of what Habermas has termed the *technical* perspective. This is the perspective characterizing the empirical and analytical sciences, the basic stance taken whether the realm of inquiry is the physical, biological or human world.

What is the outcome of all these technical choices, with their rigorous definitions, clear methodologies, maps, models and equations? All of them lead to the creation of *texts*, where we must think of 'texts' not simply as old documents or examples of literature, but things constructed out of the technical perspective to be given eventually human interpretation and meaning. To give *meaning* to a text is to interpret it, and we have reached here the second perspective of Habermas, the *hermeneutic* perspective. The act of interpretation is a second, critical step, one in which we must bring to bear every scrap of knowledge, insight and imagination we possess to inform the task. And if our task is to create that intersubjectively shared knowledge we call science, we must be able to convince others that both the text created out of the technical perspective, and the interpretation giving meaning from the hermeneutic perspective, are valid. This means they represent *a* truth – at least for the historic, contingent moment – *a* truth that is always subject to reinterpretation when seen from the wisdom of later hindsight. In essence, as geographers we are always trying to tell a persuasive story, one that will convince others, and stand up to their rigorous and perhaps sceptical scrutiny. Whether the story is told quietly, or infused with a sense of strident urgency, all convincing descriptions lie within *rhetoric* – the old, and in its original meaning thoroughly honourable, art of persuasion.

So science, and geography conducted in this tradition, is ultimately storytelling? Storytelling based upon convincing interpretations of text? Yes, of course: what else can it be, given that the world is being shaped by human beings in all their cultural variety? The historical record lets us see how the 'texts' of one time are reinterpreted at the next, or discarded or forgotten altogether for other texts considered more convincing. For example, the mathematical text created from the technical perspective by Einstein, and his extraordinary interpretation of that text (only slowly and sometimes reluctantly understood), absorbed the earlier story of Newton that had stood for 300 years. In Copenhagen, during the heydays of Niels Bohr, atomic models or texts were created almost weekly, interpreted with excitement and joy, only to fall and be discarded as a more convincing text came along. Creating texts and interpreting them convincingly is a very difficult business, and it always seems much easier to destroy with criticism than create with conviction. Sometimes it seems that most practitioners are sitting in the grandstands of the Coliseum, holding their theoretical thumbs down in condescending disapproval, rather than being willing to come down into the arena of empirical investigation to be soiled with the blood and the sand. In time, we are all wrong, provided we have created something worth examining for its truth in the first place.

Through technical text creation and hermeneutic interpretation, rhetoric seeks to gain the assent of the reader. But technical choices and interpretations may differ, because as human acts neither takes place in a void. Both are informed and shaped by the examined or unexamined values of those who inquire, and critical self-reflection can only expose, and perhaps modify, the shaping values, not cause them to disappear. Anne Buttimer has pointed again and again to the crucial role of values underlying and shaping inquiry, and she has greatly heightened the awareness of geographers to the last of Habermas's perspectives, the *emancipatory*. This is precisely the self-reflective concern to examine the act of inquiry itself, although we must be a little careful here. Taken to an extreme, the emancipatory perspective implies, once again, that we can stand outside of history, and detach ourselves from the world confident that we know its ultimate purpose and trajectory. To put it politely, this, I think, is nonsense. But taken thoughtfully, the emancipatory perspective bids us ask how we might alter the social and spatial structures to create a more decent and humane world. What we should never lose sight of is that those *structures* are either defined, created and made operational by us out of the technical perspective, or they remain the buzz words that are so liberally sprinkled over the discourse of social theory today. Suppose we make inappropriate choices, and create inappropriate texts? Then our interpretative acts, informed by our values, will tell the wrong story, or none at all. So back we circle to try again, to try to get the story right, to create the always growing, always changing story that *is* modern geography.

Anne Buttimer
University of Lund-Clark University
1938–

It is not a vicious circle, one that leads us back despairingly to the same beginning, but a hermeneutic circle that takes us back to new starting points that differ from those we started from before precisely because we failed the first time around. The history of any good science is the history of thoughtful failures.

Part VIII

Geography in the future

International geography: strengthening the fabric

27

Whether you have just dipped in here and there to browse, or whether you have carefully read every word, it must have struck you that there is a definite slant to this book. As I noted in the Preface, I have only had time to focus on a few of the new developments in geography, simply because these are the parts of the story I want to tell. There is also a fine tradition of humanistic learning and writing, particularly in the area of historical geography, but I feel this is a continuation of past strengths, rather than the generation of new perspectives, and there are others who are far more competent to tell that tale. But you must also be aware of another distinct bias if you have got this far: nearly all of the examples we have looked at are by geographers from the United States, Canada, Britain and Sweden. Is this fair? Are the new ideas to be found almost exclusively in these four countries – what the French call *les Anglo-Saxons*? Is this where the explosion in modern geography started, and continues to gather its strength?

Now I am probably going to make myself extremely unpopular around the world, and I shall undoubtedly be accused of nationalistic chauvinism, ignorance, insensitivity, arrogance, bias . . . and a host of other rather unpleasant attributes, but I am going to answer *Yes* to those three questions, with a bit of a *caveat* here and there – which is the Latin word for a loophole you can wriggle out of if you have to. In the first place, I really *have* to answer Yes, because I call myself a Professor, and I feel a certain amount of moral obligation to tell the truth as I see it. But I am also going to answer Yes, because it is going to be fun to see if anyone can marshall reasonably strong evidence that I am wrong. With a few exceptions here and there, virtually all the running has been made in these four countries, and that means that nearly all the major theoretical ideas and useful applications have appeared in this rather small set. It may be that there are vast and important bodies of geographic theory published in Hungarian and Chinese, and my ignorance of these languages means that I am totally cut off from them. In that case, this book has totally misled you, and you really ought to start learning Chinese, and after that Hungarian.

Now underneath that apparently facetious invitation to learn Hungarian and Chinese, there is actually something very important to point to and think about. Ideas, and this is in essence what geography is about, are expressed and understood in language, and to my mind language is both the great glory of humanity and its most awful curse. It is a great glory because we are, in a deep sense, our language (Chapter 25), our language really is us, and without it we are not. Paradoxically, it is also our terrible curse, for it forms the greatest of barriers between us, and barriers always mean broken, or very weak, connections. Weak connections mean weak and fragmented structures, and fragmented structures mean that there are great gaps and obstructions to any ideas that might be transmitted on them. It is a sad fact that geography, as a world-wide discipline, exists in a number of pieces, often with distressingly little connection between them.

These breaks in the structure, these rents and tears in the fabric of the discipline, mean that new ideas, new ways of approaching problems, and new ways of seeing problems are likely to be transmitted very slowly, unless they are published in a language that is accessible. Today that means English, and I say this not with a sense of pride, but rather a sense of sadness. It is an acknowledgment of a simple fact, not a cause for anyone's rejoicing. This linguistic division of the world has meant, in general, a very slow diffusion of the exciting ideas that have changed the face of geography, ideas which continue to be developed, opened up and extended. This means that in a number of countries of Europe, Latin America, Africa and Asia, geography is about twenty years behind the times, with its students only now going through some of the experimental stages that characterized the more *avant-garde* departments in North America, Britain and Sweden in the late 1950s and early 1960s. In the meantime, geography has continued to develop very rapidly, both in theoretical and empirical-applied directions, so the frontiers of research sometimes seem increasingly difficult to reach.

The inclusion of Sweden as a major generator of geographic

theory and applications is significant: most Swedes in the universities speak English (you can lecture slowly and clearly to a first-year student audience and have 90 per cent comprehension – perhaps even higher than native English-speaking students), and many research publications in Sweden are published in English. Swedish ideas flow relatively easily on the international structure – and, of course, ideas in English flow easily the other way. In this respect, the *Lund Studies in Human Geography* have been particularly important, and it is these monographs that have introduced to the wider world the work of many Swedes and Finns – a number of whom we have met in this book. The series has also provided an important opportunity for British and American geographers to publish their work, some of which was blocked by entrenched Establishment attitudes in the early days of the (R)evolution. Occasionally a study will be published in the Lund series in French, but the ideas seem to be slowed down considerably – even to the French. For example, a quarter of a century after its publication in French, I can find no reference in the French geography literature to the early, and now classic study by Edgar Kant on the central places of his native Estonia (Chapter 9). A tremendous barrier exists between the French- and English-speaking worlds, and even the references used by one of the most *avant-garde* groups of French geographers are twice as old on the average as their English counterparts, a sad and direct comment of the way language slows up developing ideas. The French, of course, can retort that few publications in French are referenced by the English-speaking world, but geography in France has been extremely conservative, inbred, and dominated by Paris (much to the frustration of a number of young geographers), and from the standpoint of the English-speaking world there is little to reference in terms of genuinely fresh and new ideas. Once again, it seems to be a matter of catching up, rather than making the running.

The problem, as always, is that when new ideas come along there is always an older body of entrenched opinion resisting them. In a very large country like the United States, it is difficult to stop ideas for very long. As we saw in Chapter 3, new journals spring up to by-pass the gatekeeping of the old, and the tradition of publishing discussion papers is strong. But in a small country, with only a few universities, and an older generation holding to a conservative tradition, things can become much more difficult. Yet here and there a few people can start to open things up, exposing students to new ideas, pointing to new journals, new possibilities, and other more

Jorge Gaspar
University of Lisbon
1942–

theoretical and practically oriented traditions. For example, in Portugal, Jorge Gaspar represents an important connection to English and Swedish ideas (he was a visitor at Lund for many months), and through him students at Lisbon have been exposed to new methodologies and problems. He has also helped a geographic planning group in the Regional Planning Commission at Coimbra, for the idea that geographers might contribute their expertise to urban, regional and national planning did not arrive with the French school that has traditionally dominated Portuguese geography. In France, it is the economists and engineers who do the planning, often with rather technically oriented and insensitive results.

It is the same in Spain, where Horacio Capel at Barcelona (always a rather rebellious city of Catalonia) has exposed his students to new perspectives, while skilfully keeping a sense of the best of traditional ways. It seems that his own deep concern for the intellectual history of geography has enabled him to stand open continually to new advances, greatly to the benefit of those around him. In Brazil, Antonio Christofelletti has tried to open up a very conservative tradition, founding a new journal with a theoretical orientation, and encouraging young Brazilian geographers to apply their knowledge and expertise. In Belgium, Hubert Beguin has shaped a new generation with his own theoretical concern; and in France Paul Claval has undertaken an invaluable service by synthesizing, in a series of books, a large research literature in English for a French student audience. As for Dutch geographers, like Christian van Paassen and Joost Hauer, they have always had much closer connections to the English and Swedish worlds, and they have shared in, and contributed to, the much more applied and practical tradition of spatial planning.

Horacio Capel
University of Barcelona
1941–

In Japan, geography was established at the turn of the century at Tokyo and Kyoto, with the groundwork in theoretical geography being laid by early pioneers such as Isamu Matsui and Joji Ezawa just before the Second World War. Today, there is a strongly applied tradition, and many geographers work closely with colleagues in engineering, economics and sociology. The same practical and applied tradition has also been extremely strong in Poland, where enormous problems of spatial reconstruction – new roads, railways, urban areas, border changes, etc. – had to be faced after the war. Both Kazimierz Dziewonski and Stanislaw Leszczycki were active in encouraging the application of new methods, and Antoni Kuklinski has done much to further the contacts with other geographers overseas.

Antonio Christofelletti
Paulist State University of Brazil
1936–

I have deliberately tried to give you a sense of a few of the individual faces, because what we always see are ideas moving through *people* who are themselves connected to a larger and rapidly developing body of ideas. Geography, nationally or internationally, is not a vast, faceless and anonymous 'thing', but a highly dynamic structure made up of real people entering the discipline, retiring from it, and changing and growing as they develop intellectually. Perhaps we can carry out a little thought experiment to make this idea of individual connections and intellectual structures, in a sense the 'fabric' of international geography, more concrete. Suppose we take the set of all professional geographers in the world as the rows of a big matrix, and all the books and articles and monographs they refer to in their writing and research as the columns (Figure 27.1). Suppose we put an asterisk * if a person uses a particular

Hubert Beguin
Université Catholique de Louvain
1932–

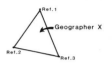

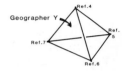

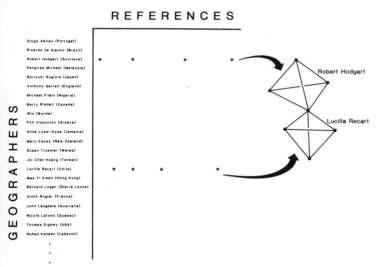

Figure 27.1: Geographers connected by the references they use.

Paul Claval
University of Paris

reference, and we think of this reference as a point or vertex of a polyhedron. This means that we can represent someone using three references as a triangle, another using four references as a tetrahedron, and so on. If we do not confine our imaginations to the polyhedra we can draw on this page, we can go to any number of dimensions we like. A very eclectic geographer, referring to many things, may be high-dimensional; another, working in a very specialized field, may be quite low-dimensional. Notice that if two geographer-polyhedra happen to use the same reference they will be connected together, and it is the connections that make structures. Of course, this is not

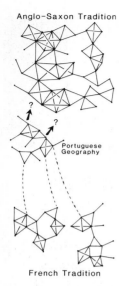

Anglo-Saxon Tradition

Portuguese
Geography

French Tradition

Christian van Paassen
University of Amsterdam
1917–

the only way we could operationalize the intuitive notion of structure. We might just take the set of geographers, and examine how they are connected by personal acquaintances, or the strength of their research interests, and so on.

So for this particular thought experiment, I would like you to try to visualize all the geographers around the world as polyhedra of varying dimensionality, all connected up in varying degrees by the research references they share. Using this rather strange, geometric but useful way of thinking, it is clear that people working closely together in the same area, say diffusion studies (where an actual analysis like this has been made), will be closely connected together, and perhaps disconnected entirely, or connected only through a long chain, to others working in a totally different area. Whole traditions of geography, say the French and Anglo-Saxon, might be very tenuously connected, but who are the bilingual polyhedra making up even that small amount of connective tissue? I have the distinct impression, for example, that the Portuguese polyhedra are disconnecting more and more from the former, and adding connections to the latter. That Anglo-Saxon piece of the structure is very big and getting bigger all the time, because all the Australian, New Zealand, Indian, Nigerian, Sierra Leoneian, Jamaican, Malaysian, Pakistani . . . geographers are there too. The more the connections, the stronger the structure, and the stronger the structure the easier it is for high-dimensional ideas to exist and move from one person-polyhedron to another. And we must not forget that if our geographer-polyhedra are professors, then research students are continually attaching themselves to certain faces (certain combinations of research interests), only to grow and become important parts of the structure themselves.

This strange way of seeing gives us other insights: the structure is not static at all, but constantly changing as new connections are made, as old geographers retire and die and are crushed out of existence (what else is dying?), and as new geographers arrive on the professional scene. It is a strange structure, one in which the people create the references, which then form new possibilities for connection. Notice that if we look the other way, down the columns of the matrix, we can represent the references themselves as polyhedra, this time with people as the vertices. Which are the high-dimensional references that form such important intellectual and connective tissue in geography? And just think how terribly important it is that these are recognized quickly and translated! If you do not read Hungarian, how can you use the new insights from

336

that brilliant article to raise your own dimensionality and strengthen the structure of geography?

So when we talk about geography as an international enterprise (or even something as small and local as a departmental enterprise in a university), we are really talking about *structural change*, and asking what we can do to strengthen the fabric, and make it higher-dimensional and better connected, so that higher-dimensional ideas can live and move on it. Thinking in this geometric framework gives us a clue to possible ways of changing the structure. Firstly, we can increase the sets of people and references, but this will work only if the people have the capacity to be high-dimensional, and the new references are important enough to be actually referred to. Otherwise, we could end up with a more fragmented structure than we started with. Every field needs high-dimensional people and high-dimensional ideas.

Joost Hauer
State University of Utrecht

Now one of the best ways of creating connections is to let people meet each other face to face so they can discuss the ideas and learn from each other directly. This is particularly important for young geographers close to the thresholds of their careers, for personal contacts forged early on often lead to strong and life-long connections. Unfortunately, the national 'Establishments' invariably devote most of the available travel funds to themselves, the jet-setting 'Call Girls' we met before who have put in their time on the national and international committees, and have finally got their hands on the purse strings. Very seldom does anything of genuine intellectual advance come from the international conferences and meetings of such people. The problem is that very few funds are available for young scholars, even though they will soon form the major part of the fabric of geography in the future. Much could be done if some genuinely altruistic older people, or people in national funding bodies outside of the discipline, were to raise such issues explicitly. The young scholar, with a career still to make, generally fears to do so – for obvious reasons. Indeed, in some countries it is extremely difficult for a bright young geographer to accept an invitation to lecture, or attend a conference, overseas, unless the old Herr Doktor Professor and all the *Dozents* of his or her department have been invited first. Otherwise the career of the young geographer may be severely jeopardized. This seems an unbelievable situation for many geographers in other countries, who feel totally free, and are only too glad to accept such opportunities for professional contact. It is unfortunately true – although an issue seldom discussed.

Joji Ezawa

337

The second way to increase the strength of the international structure is to provide funds on an annual basis for translations. National committees of geographers (under 40 – to minimize the Old Boy Effect) could review the recent literature, and pick out those works that were genuinely fresh and innovative, either illuminating a practical problem, or making an important theoretical contribution. These could be translated immediately, so making the ideas available much more quickly. In this way, national and international geographic organizations might serve the genuine intellectual tasks of geography, instead of using scarce funds for still more managerial committee meetings whose reports sink silently and unread into the abyss of time.

The day when geographers share a common language is as far away as it is for the rest of humanity, but in certain limited areas only small amounts of money could do much to strengthen the connections and the human and intellectual structures they form. It is not that the funds are not there: it is only a matter of what are seen to be the priorities. Old Call Girls today, or a young, vital and stronger geography tomorrow?

Surfing to tomorrow on time's breaking wave 28

A few years ago, the geographer Peter Haggett (Chapter 19) evoked a vivid picture of geographers as surfers, riding the crest of a wave out of the past and into the future. His image was one of excitement and daring under conditions of considerable environmental instability, a challenge worthy of the boldest and the best. The only problem was that nearly all the surfboarding geographers were facing backwards, looking towards the past, instead of out into the future. His concern, of course, was to point to the difference between trying to understand what has already happened, and trying to predict and shape that which is constantly arriving.

Since he gave geographers that rather vivid image, a number of them have turned around on their intellectual surfboards, not the least because many geographers share the growing worry about what might be arriving out of the future. Naturally, some events and conditions cannot be foreseen for any practical purposes, and there would be little we could do about them even if we knew they were coming. Despite all our attempts at earthquake prediction, involving today extremely sophisticated instruments to measure rock strains deep in the earth, earthquakes still have an unhappy propensity of arriving unexpectedly. The best predictions (perhaps a day or two) may still be those derived from watching the abnormal behaviour of animals – something the Chinese take very seriously. I have often wondered what would happen if you announced to San Franciscans that there was a 50–50 chance of a really serious earthquake arriving some time next year. Would people start moving out? Or would things go on as usual?

Other catastrophic events may be linked to disturbances on a global scale. The replacement of the cold upwelling water off the coast of Peru and Ecuador by warm tropical water (the event called 'El Niño') appears to be connected to large-scale atmospheric changes, and these may even affect the movement of high- and low-pressure zones across North America. At the moment we have no way of predicting such changes, and there would be nothing we could do about them anyway. Ordering ocean and atmospheric waves around is still as futile today as it was in the reign of King Knut.

Even in the human realm, where we may think we stand a better chance of shaping events to come, many people often have an uncomfortable sense of futility. We saw (Chapter 23) how well-intentioned efforts to bring massive supplies of food to the famine of the Sahel became mired in greed, selfishness, bureaucratic muddle and ignorance. A few years later, similar drought conditions exacerbated a growing pressure of people and animals on an already fragile environment to produce widespread starvation in Ethiopia. International efforts, including American food and Soviet transportation, tried to provide relief, but to those who deny that there is any population problem in the world, one can only pose the question, 'Then why are people there *not* producing the food they need?' Some geographers cry out for spatial equality and regional self-sufficiency, and then turn around and declare there is no population problem when people cannot feed themselves. Where does a problem begin – and end? We could point to many other areas where events seem to take their own course, where the problems of redirecting our present trajectory into the future seem so complex that we feel impotent in the face of them.

And yet . . . and yet despite the complexity and ignorance, and the sense of discouragement often felt in the face of imponderable problems, many who see the world through the eyes of the geographer are becoming increasingly involved, are trying to help (each in their own and individual way), are turning around on those surfboards and shifting their weight in an attempt to steer away from the coral reefs ahead. *Time*, of course, will always bring us the unexpected; we are humans not gods, and peer as we might through the obscuring spray of the present, we cannot, any of us, see very far ahead. On the other hand, the overwhelming, though by no means exclusive, characteristic of the geographic perspective is not time but *space*. Not the empty wastepaper basket space that you toss things into, but our humanly shaped, structured and organized space. This is the geographic space shaped by human meaning, where today we see the traces of our own past humanity, the same spatial traces that will shape, in turn, tomorrow. Societies moving through time are in constant 'dialogue' with the geographic space they shaped in the past, the same geographic space that forms the setting today, and the one that will shape them in turn as they move towards the future. The human

340

organization of geographic space has an historical tenacity that allows us to change it only relatively slowly. Apart from a devastating atomic war, or a large comet roaring out of the darkness to collide with this little blue and white globe we call our home, the geographic map tomorrow will look much the same as it does today. No matter what the political regime, the ideology, or the religious convictions of the societies embedded in the historically given spatial structure. Fortunately, some things are *relatively* stable and not wholly in the realm of the unexpected. In the geographic realm, we still have a chance of shaping things slowly and thoughtfully towards a more humane and decent future.

As we have seen throughout this book, natural, living and human systems in space and time may be enormously complex, and as we move our thinking along that natural-living-human sequence the complexity increases by at least an order of magnitude each time. To state the problem in another, rather strange, but thoroughly geographical way: at one extreme we have physical systems of things whose courses and trajectories are shaped by the stable space-time geometry in which they are embedded. No one will deny that these systems are extremely complex, and it has taken some of the best thinking over the past 300 years to unravel even parts of the story. But at least the *geometry*, the underlying structure of space-time upon which the things of the physical world move and have their being, is stable – or at least so it appears to us. And since we are embedded in the space-time geometry of the universe ourselves, how could we stand outside of it and know if it were changing or not?

However, as we move along that physical-living-human continuum, the 'things' are no longer sticks and stones, but living beings who have the capacity to shape the local geometries in which they have their own being. At first not consciously, not reflectively and *thoughtfully*; but certainly instinctually, presumably through genetic programming. The coral reef, the bee hive, the ant nest and the channels of migrating birds are all marvellous worlds of living organization, but few would care to ascribe self-conscious awareness, or the capacity for thoughtful reflection, to the inhabitants. It is when we come to ourselves that self-reflective, thinking consciousness begins to shape the geometries. Not just the physical geometries, but the social, political, economic and many other geometries that allow, forbid, but do not require. We are the geometry shapers, those beings who have some capacity to alter, consciously, the multidimensional spaces in which we live out our lives. And if we are truly *thinking* beings, we must do so caringly and lovingly. Geographers, with an intellectual passion for the space and time of that speck of matter hanging there in the darkness, are deeply and caringly concerned with our planetary home. It is all we have got.

Different geographers carry out their caring in different ways: we are all different, and all have different strengths, abilities and gifts to carry out the task. Some, with a deep sense of heritage, point constantly to the landscape that has been shaped by wind, tide, flood and storm, and upon which the traces of our humanity appear as the cultivated fields, roads, buildings, bridges . . . all the works of humankind. A number of geographers are deeply involved in historical (or should it be geographical?) preservation. Others, with gifts for mathematical description, attempt to unravel and simplify the complexity, and then model it in the form of computer software, to experiment with clearly defined interactions of many variables, whose combined effects produce such huge and complex possibilities that we are becoming increasingly aware and attuned to the delicacy and fragility of the ecological and human webs. Others are teachers, and daily impart their own care and concern to students of all ages, from young children in the kindergarten to adults in night school. Their caring is conveyed in many ways: by pointing to the facts of natural systems and human pollution; to the beauties of an old urban landscape surrounding an unspoiled cathedral square; to an abstract, and perhaps mathematically stated, theoretical 'lens', through which a sense of order is seen where only chaos was perceived before. Without teachers how can new, caring and thoughtful ways of seeing become possibilities? Yet teaching is a lonely task: you never really know what the impact is at the moment, or what it may become upon thoughtful reflection in the future. But whether through teaching, writing, research or consulting, and no matter how those tasks are carried out and modulated by a variety of deeply held personal convictions, the true geographer is a person who has made a lifelong commitment to a caring concern to understand, and perhaps to shape, the physical and human spaces of our planet.

We have seen a few (and only a few) of these latest attempts in this book. They range from a concern for the dynamics of urban settlement and regional development, through the diffusion of ideas and diseases, to the modulating ideologies and philosophical reflections that shape the ways of seeing. And in this broad and eclectic field, there is always a tension of accusation: shall we be condemned for specializing in a small

342

area, knowing more and more about less and less? Or shall we butterfly our way around, lighting briefly on one flower and then another, only to be accused of intellectual superficiality? It is a tension to be lived boldly and confidently in a world where specialization is superb at taking things apart, but sometimes forgets how to put them back together again. We need, all of us, many ways of holistic seeing, providing each way illuminates yet another facet and perspective of our fascinating world.

There are many ways of seeing in geography today, and the best do illuminate the geographic world about us, and so let us say that most wonderful phrase in any human tongue, 'I never *thought* about it in that way before!' It is the old Eureka experience, the one in which something previously hidden appears, in which something concealed comes into the open clearing of our thinking for the first time to be understood. For the geographer, as for others, these moments of knowing and understanding can come at any time, often when they are least expected, and from any direction of informing and modulating ideology. Geography is full of isms today, yet each, in its own way, contains a condition of possibility for letting us see a little farther than before. Positivism, Marxism, Humanism, Pragmatism . . . Quantification and Qualification . . . they swirl around Geographia, who, in a moment of weakness or musing preoccupation, has allowed herself to be snatched up and carried away by the convictions and certainties of one way of looking. Yet 'from too close looking follows loss of sight' said the poet (Preface), and from time to time Geographia herself has to take command (Figure 28.1). There she goes, lithe and strong, totally in command at last, fed up with all the squabbles and chattering. Enough is enough. Her only concern is for true knowledge. Let Qualifactus weep for his pastoral days: in time he will dry his tears, and with an unblurred vision he will help us see in his way once again. Let Karl Marx rant and rave as El (Machismo) Barbo, stripped of all polemic and jargon, is carried helplessly through the broad Fluvius Leninus to an older demesne that floats in the great river of Time, and appears not to obey any 'laws' of history. As for Quantifactus, still clutching his jetset consultant briefcase, there is nothing to do but take him by the scruff of his neck, get him on board, and then pole him across to the bank of reality once again.

So there she goes, our magnificent Geographia, headed straight for a strange raft, with its space-time paths leading through coupling constraints to a somewhat abstract Tree of Knowledge. Watched over by the good ship *Structuration* (filled

343

Figure 28.1: Fed up with the bickering, Geographia takes command at last, carrying the Marxist revolutionary El Barbo, and the Capitalist consultant Quantifactus, back to the bank of geographical reality. The apples on the Tree of Knowledge are a bit more abstract today, but just as tempting as ever. Reprinted by permission of the anonymous artist who still enjoys oxymorons.

with hot air), the variable winds tested by a geometric kite of novel design, who can doubt that she will make it easily to the opposite bank. As for those geographers, I have the feeling that they are going to taste that fruit – even if she has to shove the apples down their throats.

Will they never learn her mysteries?

Envoi: Portrait of the author as a young geographer

Peter Gould's primordial spatial experience as a geographer occurred during the eighth week of his life, when he reached towards the light and wondered whether his cot revolved around the lamp, or the lamp went around the cot. This is a conundrum he has yet to resolve, although he has been trying to reach for the light ever since.

As a child in England, he never lived in one place for more than a year, an experience that confirmed his view that places were private, personal, and incapable of being shared, while spaces were both public and enticing because they were bound by a horizon that required one to ask what lay beyond. In 1936–7, he lived within a horizon of small villages in the Schwarzwald (Private Landscape I), stealing smoked bacon and falling off passing haycarts. Many feel that he has yet to recover from these experiences with the third dimension. Occasional forays from the mountains were made to the Bernshof near Freiburg partly to horrify a much loved great-aunt with his authentic peasant accent, and partly to receive instruction from the mathematician Ernst Zermelo on how to make the figure 8 in one continuous movement. Lessons on such fundamental properties of the integers were held during tutorial breakfasts of strong coffee and caviar.

He returned to the West Country, a necessary, though not sufficient, location to become a real geographer, where he picked blackberries, and learnt to ride a bicycle, until the outbreak of war. In 1940, he was evacuated to Cambridge, Massachusetts, an experience that so imbued him with the spirit of Lexington and Concord that he still views all monarchy as a tyranny to be overthrown as rapidly as possible. Vacations were often spent at Van Hornesville, New York (PL II), where he learnt about clearing wildernesses, the sounds a creek can

make at night, the smell of ripe apples, and the taste of water from a limestone spring.

In early 1945, he returned to England, got torpedoed on the way, but made it to shore, where he was incarcerated in one of England's oldest and finest public schools. After a year of moral neglect and amateur sadism characteristic of these institutions of higher education, he transferred to a naval college, since fathers, mothers, uncles and cousins in the navy thought it wise to send the lad to sea. Vacations, at this point in his rapidly chequering career, were spent with a colonel grandfather, who had his worst suspicions about trans-Atlantic influences confirmed when it was clear that 'the lad' knew no poetry whatsoever except 'Paul Revere's Ride', a jingle from a colonial poet whose name permanently escaped him. Breakfast, lunch, tea and garden sessions on Kipling, Swinburne, Tennyson, Browning, Byron, Shelley, Keats . . . ('but none of those damned Eliot fellers') followed, opening a path of ever-renewed delight – including some of those damned Eliot fellers.

Far from going to sea, the lad was called to geography, though not immediately in practical terms. After passing examinations in marine navigation set by the Elders of Trinity House, but failing to navigate the treacherous Latin water of Cambridge University, he joined the army in a fit of sheer bloody-mindedness for national service. Quickly learning to mould plastic explosive into beehive charges, and other manly pursuits, he was commissioned in the Gordon Highlanders and sent to the Moray Firth to train recruits. Bored to tears after a month, but sustained by weekend hikes in the Cairngorms (PL III), he made the appropriate noble and self-sacrificial noises to his regimental officers in Aberdeen, and volunteered to joint the 1st battalion in Malaya – in order to protect imperial investment in rubber and tin, and to allow that country to evolve to Westminster parliamentary democracy by the year 2050. A year of jungle patrolling, ambushes, airdrops and mosquitoes was relieved by six weeks of rest and retraining in Singapore. This consisted of getting up at 0600 six days a week to practice highland fling steps under the eyes of the colonel and pipemajor, and attending regimental dinners at long tables loaded down with silver captured from the French after various skirmishes such as Salamanca and Waterloo. He has loved Scottish (and all other forms) of dancing ever since, demonstrating how a rigorous military training can enhance an appreciation for the Arts.

After the army, he spent some months at his postwar home in Salcombe, South Devon (PL IV), walking the wild cliffs, watching birds, and messing around in boats. He then returned to central New York as a student at Colgate University, continuing to Northwestern for graduate work. A year of fieldwork in Ghana (PL V), his first love in Africa, followed, leading to a teaching position in 1960. A further year of fieldwork followed in Tanzania, living in the shadow of Kilimanjaro (PL VI), at which point he joined the Pennsylvania State University.

He has remained among the beautiful hills and valleys of central Pennsylvania (PL VII), except for sabbatical years in Skåne (PL VIII) and the Vercors (PL IX); consulting trips to Portugal, to savour a gentle November light on the resting landscape of Coimbra (PL X); and occasional journeys to London, Cambridge, Paris, Strasbourg, Copenhagen, Lund, Kyoto, Port of Spain, Cairo, Lisbon, Caracas, Mexico City, Kingston, Bellagio, Abisko and Reykjavik – all justified on the grounds of selfless service to the human family, deep professional dedication, and a pure and holy desire to further knowledge. The fact is, he seizes every opportunity to travel, and does so with unbridled hedonistic delight.

The author of a dozen books, and more than 100 articles and essays in over 50 journals and 40 collections, he has served as consulting editor for half a dozen professional journals, and as a visiting lecturer at more than 50 universities. He holds the degrees BA (Colgate), MA and PhD (Northwestern), and DSc (Strasbourg).

Peter Gould
The Pennsylvania State University
1932–

Index

Index